Antarctic Wildlife

An Antarctic Cormorant, also known as the Blue-eyed Shag, feeding its young on the nest.

A Southern Giant Petrel glides effortlessly over the ocean.

A mixed company of Royal and King Penguins on the shore at Macquarie Island.

Portrait of an immature Southern Elephant Seal.

Overleaf *The* Lindblad Explorer *at Port Lockroy with Antarctic Cormorants in the foreground.*

Antarctic Wildlife

PHOTOGRAPHS BY **Eric Hosking**

TEXT BY **Bryan Sage**

Facts On File, Inc.
460 Park Avenue South
New York, N.Y. 10016

Top *A Rockhopper Penguin looks curiously down from its vantage point on a rock.*

Centre *In this shot of a Rockhopper Penguin the structure of the crest on the head is shown to advantage.*

Bottom *A Gentoo Penguin using flippers and tail to maintain balance.*

To Lars Eric Linblad who made it all possible.

Antarctic Wildlife

First published in the United Kingdom in 1982 by Croom Helm Ltd, 2–10 St John's Road, London SW11.

First published in the United States of America in 1982 by Facts On File, Inc.

Library of Congress Cataloging in Publication Data

Hosking, Eric John.
Antarctic wildlife.
Includes index.
1. Zoology – Antarctic regions. 2. Zoology – Antarctic regions – Pictorial works. I. Sage, Bryan L. II. Title.
QL106.H67 591.998'9 82-1382
ISBN 0-87196-679-4 AACR2

Designed by Tom Deas

Typeset by TJB Photosetting, South Witham, Lincs.
Printed and bound in Great Britain by
Hazell Watson & Viney Ltd., Aylesbury, Bucks.

Contents

Top *Portrait of a Chinstrap Penguin showing the facial pattern which gives the bird its name.*

Centre *A pair of Chinstrap Penguins with their two young.*

Bottom *Adult Chinstrap Penguin with young.*

Foreword *Keith Shackleton, President of the Society of Wildlife Artists 6*

Acknowledgements *8*

Introduction *10*

Map of the Region

1 The Historical Background *18*

2 The Antarctic and Its Ecology *26*

The Antarctic Continent *32*

The Maritime Antarctic *37*

The Subantarctic *42*

3 Penguins *52*

4 Albatrosses *92*

5 Other Birds *108*

6 Seals *128*

7 Wildlife Photography in the Antarctic *142*

Index *158*

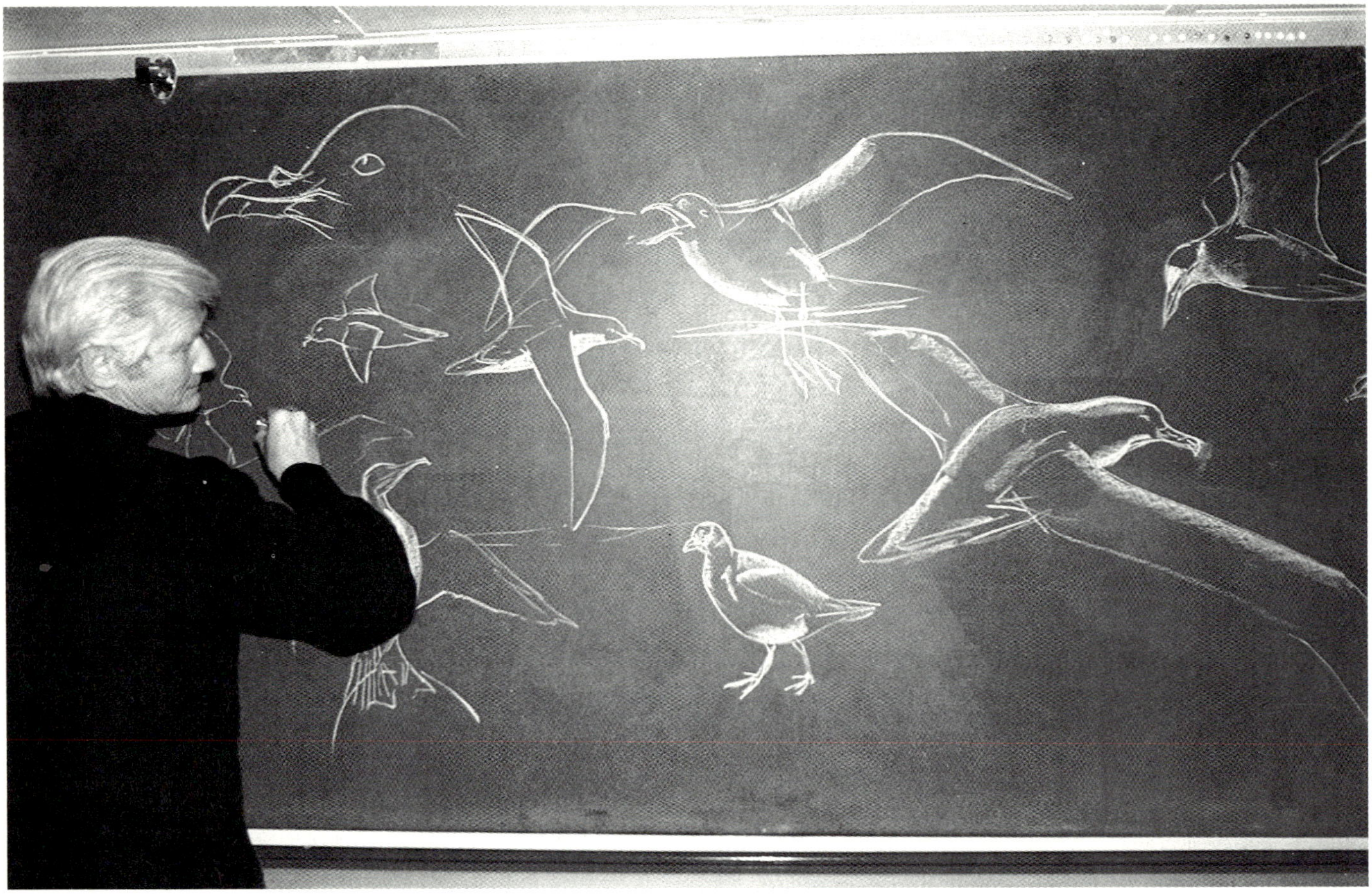

Above left *Keith Shackleton demonstrates his artistic skills in the lecture room on the* Lindblad Explorer.

Above right *Rough seas break over the bows of the* Lindblad Explorer.

Overleaf *A Blue Petrel flies over the calm surface of an ocean lit by the dying rays of the sun.*

Foreword

There has always been an indefinable love/hate relationship between painting and photography. It stems I suppose, from the results of both endeavours being two-dimensional, pictorial interpretations of reality, that may be appraised on somewhat similar terms. For all that, 'photographic' has become one of the less complimentary adjectives in the art critic's armoury.

Painters can learn much from photographers and I believe the reverse is true; the practitioners of each can certainly live happily together, especially when they share a very deep love for the same subject. Total reconciliation may sometimes be elusive but enjoyable sparks of argument are struck all along the way. For me, the ability to take good pictures with a camera, to play the piano, to read a balance sheet, to knit a complicated sweater, have always basked in an admiration close to awe.

Eric Hosking was the first real 'Rembrandt of the Shutter' I ever met. I was a schoolboy and he was already acclaimed one of the world's leading wildlife photographers. We have been friends ever since. He was always an ardent collector of paintings and drawings too, so we had the chance over the years to argue the brush versus view-finder bit to the point of exhaustion. Cameras and techniques changed and he embraced each breakthrough with no hint of remorse, that photography was becoming

more fool-proof by the minute, more popular, broader based and harder than ever for an old hand to maintain supremacy. There were after all new places and new subjects. 43 years after our first meeting I found myself roping him to the rolling, pitching rail-stanchions of the MS *Lindblad Explorer* so that he could relax free-handed and swing his camera in tune with our albatross escort. Eric was going South to the ice and this book is the happy outcome.

Life has few joys to equal the sharing of a long-held enthusiasm with another.

Eric is an exact contemporary of Captain Scott's son, so he grew up in the years when parents read to their children of polar exploits that kindled the imagination, put thoughts into perspective and established values. The tragedy of the *Terra Nova* expedition with its story of human courage, endurance and sacrifice proudly underpinned the endeavours of generations to follow. Despite the efforts of professional denegrators, it still does. For Eric Hosking, to see Antarctica and especially the historic heartland of Ross Island had been a life's ambition tucked away in patience with many others. The difference was that it topped the priority list and therefore had to happen.

'Seeing' for Eric, means photography. 'I've just emerged from six weeks in my darkroom', he once told me, 'and I'd no idea how much Dorothy and I saw in Kenya'.

This time it was different. I even found him climbing a small mountain with no camera at all – because the devastating beauty of this place demands that on occasion one must simply look and pay homage and savour the privilege.

Antarctica is a place where photographers and painters may see more closely through the same eyes than anywhere else I know. The landscape is savage and pure and intrinsic. Every shape, each line is a natural happening, a home-grown abstract. Against all this are the wild animals that suggest their evolution contained a purely aesthetic demand to compliment their harsh environment as strong as the needs of survival itself.

These are the scenes Eric Hosking has worked on and having seen the camera aimed and clicked on so many occasions, it has been hard to contain the mounting impatience to see the results of an expertise that has covered so much of the world's wilderness, finally let loose in the greatest of them all.

Now it is here, those who see this book could not help but be inspired by the visual wonders of the land. Bryan Sage's excellent profile fills in the background story of exploration and discovery, climate and geomorphology as well as the main task of presenting an up-to-date picture of the fauna and flora of Antarctica and the sub-Antarctic islands of the Southern Oceans. My personal hope is that such enjoyment will grow from individual wonderment into an insistence by all nations that this last continent be kept inviolate. Up until now human intrusion has exercised an impact upon the environment of little more consequence than a landing on the moon, but it has brought us two things – scientific facts and human inspiration. We are the richer for both.

The future will inevitably increase the yields of these harvests and in the gathering one hopes, will come a realisation that our planet needs a beautiful and unsullied continent as never before. It needs it as a living laboratory for scientific study on the one hand – on the other as an escape and to nourish the spiritual needs of mankind.

Keith Shackleton
MS *Lindblad Explorer*

Acknowledgements

The photographs in this book were taken on two wonderful expeditions to the Antarctic in the *Lindblad Explorer*. The first of these was to the Northern Peninsular, that arm that sticks out from the main land mass like an inverted comma. During the second cruise we circumnavigated the Antarctic boarding the ship at Ushuaia in Tierra del Fuego, the southernmost town in the world, and leaving it at Stewart Island in the south of New Zealand.

Although I would like to express my thanks to very many friends, space allows me to mention only a few by name. My thanks go first to Nigel Sitwell who suggested to Lars Eric Lindblad (to whom this book is dedicated) that I might help the tourists with their photography by giving talks on how to get the best out of their cameras and to help with the identification of all the birds and mammals we saw. All the colour photographs were taken on Olympus OM-2n cameras with a range of lenses from 18 mm super-wide angle to 400 mm telephoto and I am indebted to Barry Taylor of the Olympus Optical Co. and his staff for all the assistance they gave me.

The captain, Hasse Nilsson, and crew of the *Lindblad Explorer* could not help enough, the ship often being diverted slowly to a place where seals were asleep on ice floes, or towards a pod of huge Fin Whales, the second largest mammal in the world, or to a magnificent sculptured iceberg which we would cruise right round to see it on all sides and to take our photographs. Once I was foolish enough to fall into the Antarctic sea but Mike McDowell, our extremely competent tour manager, well over six foot tall and as strong as an ox, fished me out before I had submerged. Unfortunately my rucksack filled with sea-water and a valuable Olympus camera, a motor drive and two lenses were ruined by salt water.

In 1946 Colonel Niall Rankin made an expedition to South Georgia, taking his own boat (a converted Royal National Lifeboat) with him as well as his brass bound teak camera with glass plates measuring 6½ by 4¾ inches. How different from the modern 35 mm! After his sudden death in Africa I was asked to look after his photographs and I am delighted to be able to include just a few of them in this book. They are Wandering Albatross (p. 98), Light-mantled Sooty Albatross (p. 102), Grey-headed Albatross (p. 103), Macaroni Penguins (p. 76) and King Penguins (p. 49).

Whenever we think of Emperor Penguins we visualise them in Antarctic surroundings. They are the only birds that breed in the horrendous weather of the Antarctic winter and

because I had not been there at this time I am indebted to Peter Prince and Bruce Pearson for loaning us two of their transparencies (p. 90 and 91).

As I have said elsewhere I am a photographer not an author so I want to express my very grateful thanks to Bryan Sage who has written the text. He has a passion and a love of the Antarctic that goes back to his childhood days.

My friend, Keith Shackleton, was down there with me. He is one of our finest wildlife artists, a pilot, yachtsman and an excellent speaker and he has kindly written the Foreword. It was Keith who thought of the idea of lashing me to the rail of the ship during heavy seas so that I could concentrate on the photography without worrying about keeping my balance.

On the two trips I took some 15,000 photographs so it was no easy task to select those to go in this book and I am grateful to Tom Deas for helping with this and for so skilfully and attractively laying them out. Jo Hemmings, the Natural History Editor of Croom Helm, was full of enthusiasm and it was a tonic to work with her. Christopher Helm had such faith in me he drew up an agreement for this book and signed it before I had even been on the second expedition and for this I am deeply grateful.

David, my younger son who is now in partnership with me, shouldered much of the routine business to leave me free to deal with this book. Finally, and most importantly, I owe more to Dorothy, my wife, than perhaps most husbands do. Now that our children are all grown up she accompanies me on all our major expeditions, doing all the donkey work (packing, letter writing, finance, preparing food and drink) and carrying extra cameras, lenses, spare film and so on, leaving me free to be totally absorbed in the photography. Many of the photographs in this book would never have been taken if she had not been in the background knowing just what lens I would need almost before I did.

The Antarctic is an unbelievable wonderland, one of the very few unspoilt lands left in our world – long may it remain so.

Eric Hosking

0°
Zavodovski Island
Vindication Island
South Georgia
South Sandwich Islands
Montagu Island
Cook Island
Scotia Arc
Falkland Islands
South Orkney Islands
Coronation Island
Laurie Island
Signy Island
Elephant Island
Gerlache Strait
King George Island
Snow Hill Island
South America
Deception Island
South Shetland Islands
Palmer Archipelago
Anvers Island
Palmer Station
Princess Martha Land
Queen Maud Land
Weddell Sea
Theron Mountains
Mount Jackson
Mount Faraway
Marguerite Bay
80°
Ronne Ice Shelf
Bellingshausen Sea
Sentinel Range
Vinson Massif
South Pole
Lesser Antarctica
Greater Antarctica
Horlick Mountains
Dufek Intrusion
Byrd Station
Queen Maud Mountains
Vostock Station
Transantarctic Mountains
Mount Hamilton
Bay of Whales
Mount Erebus
Ross Island
Beaufort Island
MacMurdo Sound
Ross Sea
Victoria Land
Ile de la Possession
Cape Adare
Possession Island
Balleny Islands
Scott Island
60°
180°
Hurd Point
Macquarie Islan

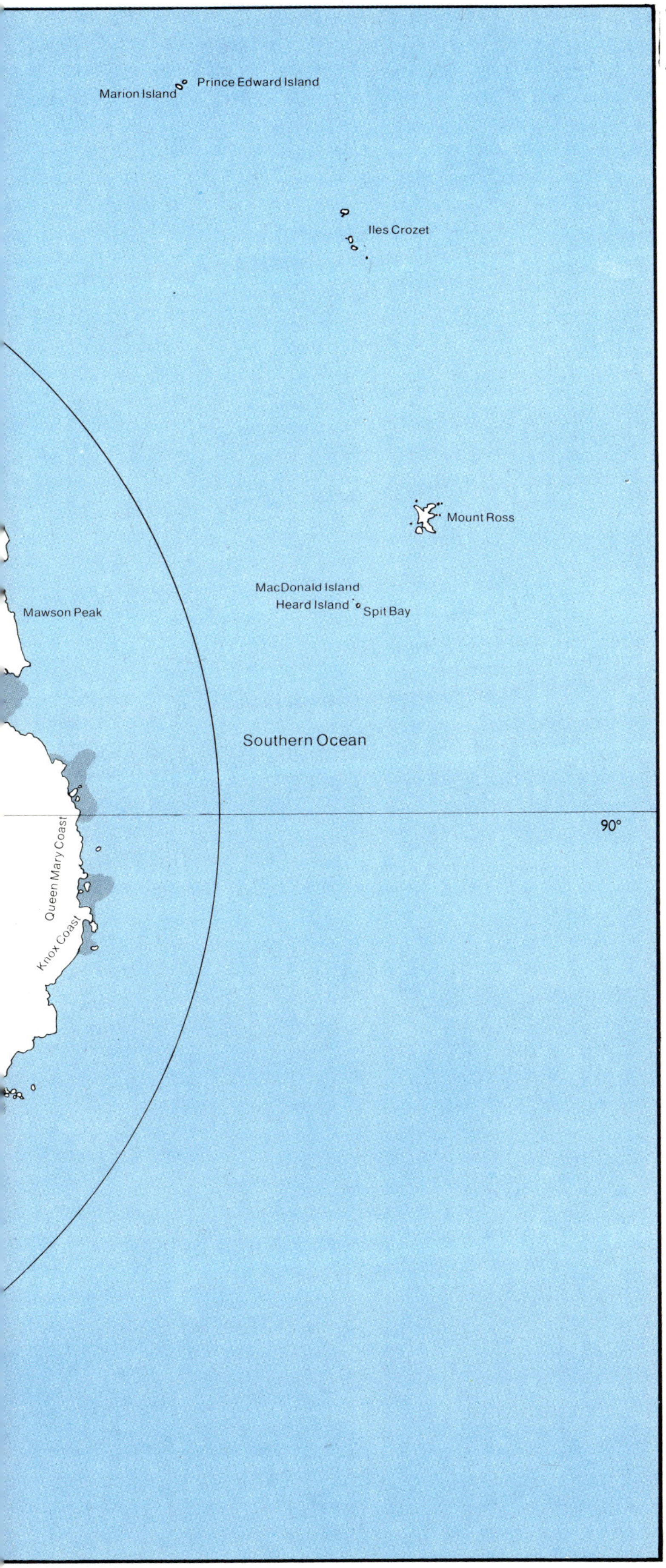

Introduction

Some 1500 years ago Greek scholars, convinced that they inhabited a globe, surmised the existence of an icebound antarctic region. With the knowledge that the climate became steadily colder as the North Pole was approached, they concluded that so it must also become progressively colder towards the South Pole. Many centuries were to elapse before these predictions were to be proved correct. A thousand years or so were to pass before man had developed navigational techniques to the point where he could contemplate exploring the great expanse of the extreme southern hemisphere.

Maps of the world compiled in the sixteenth and seventeenth centuries showed a continent surrounding the south geographic pole. The early cartographers believed that this landmass filled the polar region and was linked with the southern extremities of Africa and the Americas. They called it *terra australis incognita*, and it was an entirely imaginary landmass that no one had ever seen. So it was, that long after the spreading tide of humanity had invaded all other lands of the earth, Antarctica remained a mystery. Guarded by its apparently impenetrable ring of pack-ice it remained unknown until it was first glimpsed by man about 1820. At that time one or two intrepid mariners saw on the southern horizon the pinnacles of a distant land, later to be known as Antarctica, the last (and fifth largest) of the world's seven continents to be discovered and penetrated by man. As we shall later see, the ice barrier guarding this great continent was first penetrated by Captain James Clark Ross in 1841, but even so he was unable to land on the continent itself.

Although this is primarily a book about the wildlife of the Antarctic, the human history of the area is also briefly reviewed. Man has had, and will undoubtedly continue to have, a considerable impact on the ecology of the southern end of the

globe, so he too must be considered. First he decimated the populations of seals and whales. Now the threats include increased tourism, the large-scale harvesting of krill, and possibly at some time in the future the exploitation of oil. The harvesting of krill may have far-reaching effects. The antarctic marine ecosystem is simple in that krill, directly or indirectly, constitutes the primary food base for whales, seals, birds, pelagic and demersal fishes, and larger pelagic invertebrates, principally squid. If any part of this system is knocked off balance then, in theory, the stability of the whole system could be jeopardised.

If the scope of this book were confined to the wildlife of the Antarctic Continent itself, it would be difficult to find much to say that would be of any great interest to the general reader. The reason for this is very simple, for the continent has no land mammals, reptiles, amphibians or freshwater fish. The dominant land animals are arthropods, mites, and insects and their relatives, and a good many of these can be seen only with the aid of a microscope. It may also be added that there are no flowering plants on the continent. For these reasons the book covers the Antarctic as a whole, and it is necessary to define what is meant by this term in the present context.

Maps of the Antarctic show the Antarctic Circle at 66°33′S, but this is an abstract astronomer's boundary that is of little value from an ecological point of view. The boundary of the Antarctic is, therefore, taken to be the Antarctic Convergence which runs irregularly between 50°S and 60°S. We cannot *see* this boundary either, but it can very easily be detected since it marks the border between one oceanic environment and another, as evidenced by a marked change in water temperature. The Antarctic Convergence is the line at which the cold and less

Above: *McMurdo Sound showing Scott's hut from the* Discovery *Expedition, the* Lindblad Explorer *and the present United States base.*

Right: *Cape Hallett – the bleak home of South Polar skuas.*

MS World Discoverer *in an ice pack off of Deception Island.*

The lighting in this picture gives a 'sugar icing' effect to these icebergs.

saline water flowing north from the Antarctic Continent sinks beneath the warmer and saltier water coming from the north. With this definition the Antarctic thus includes not only the continent itself, but also a considerable number of islands of varying character.

The inclusion of these islands considerably extends the range of animals that we can discuss. The sea south of the Antarctic Convergence covers an area of nearly 36 million square kilometres, the pack-ice belt at its maximum extent in late winter covers 22 million square kilometres or 60 per cent of the sea, and at minimum about 4 million square kilometres or 11 per cent. The book does not, however, attempt to deal with any of the marine animals of the southern oceans except the seals. Seals, like seabirds, have to come to land to breed, and so fall within the orbit of this book. Nor have I attempted to deal with animals that are alien to the Antarctic but have been introduced there by man, as for example a New Zealand bird called the Weka *Gallirallus australis* that was introduced to Macquarie Island, or the reindeer *Rangifer tarandus* introduced to South Georgia.

There is still much that we do not know about the Antarctic Continent. Most of the surface has been mapped in the course of surveys over the past quarter of a century or so, and the thickness of the ice has been measured. As a result of the application of sophisticated seismic sounding and radiosonde techniques, the *shape* of the continent beneath the great ice sheet is now known with considerable accuracy. However, the actual *nature* of the rocks beneath the ice is still very much a matter of speculation and discussion. Interesting discoveries continue to be made each year. By using a unique variety of detection systems, geologists have now located one of the

A beautifully eroded iceberg resembling a medieval castle.

largest complexes of mineral-bearing rock known in the world. Called the Dufek intrusion, it covers about 50,000 square kilometres between the South Pole and the Weddell Sea, and consists of a great upwelling of magnetic magma which occurred when Antarctica was edging away from Africa 170 million years ago.

For the vast majority of people the Antarctic Continent will remain a land of mystery known only through the medium of the written word, or occasionally brought into the home by television films. Perhaps this situation will never change substantially, and possibly that is just as well. The Antarctic Continent is not a comfortable place to visit. The scientists and others who live and work there for varying periods, survive only with the aid of modern technology. It is not hard to see why this is the case, for even in the perpetual daylight of summer it is a land cold beyond the bounds of normal experience, scoured by gales and obscured by blowing snow. For any kind of life, let alone man, this is the most forbidding area on earth.

It is perhaps just as well to explain that this book is not meant to be a detailed account of the ecology of the Antarctic; that requirement is met by other very learned and comprehensive works. The text has been written to complement the excellent photographs taken by Eric Hosking. However, Chapter 2 tries to provide a reasonably detailed account of the topography of the Antarctic, and discusses also some of the interesting aspects of its ecology. It is hoped that this will prove a useful background for those unfamiliar with the area, but who may have the good fortune to find themselves at the southern end of the globe. Subsequent chapters look in more detail at some of the seals and seabirds.

There is now a great deal of international concern regarding

Twin rock towers stand sentinel in the Lemaire Channel.

the future of thc Antarctic and this has acted as an impetus to scientific research. One of the most important projects currently under way is known as BIOMASS, and is designed to provide detailed information on the biological resources of the Southern Ocean. When this research programme is completed we should have a great deal more knowledge about the ecology of this fascinating part of the world.

Finally, I must express my gratitude to various friends at the British Antarctic Survey who have been generous with help and advice, namely Dr W.N. Bonner, Dr J.P. Croxall, Dr R.M. Laws, Dr P.A. Prince and Dr R.I. Lewis Smith. I am also indebted to the librarians of the British Antarctic Survey and the Scott Polar Research Institute who have been extremely helpful in providing reference material. I am indebted, too, to Dr Terence Armstrong of the Scott Polar Research Institute who has also given me the benefit of his advice. Any errors that this book may be found to contain are, however, entirely the fault of the author.

The Historical Background

Many of the various events that make up the history of the Antarctic, and in particular that of the Antarctic Continent itself, i.e. Antarctica, are of relatively recent date. The exploration of the continent proper did not begin until after the beginning of the present century, which is still within the living memory of a good many people. One of the peculiarities of antarctic discovery and exploration is that the continent had been studied from offshore for well over half a century, and more than one nation had made territorial claims upon it before any man had actually stepped ashore. The story of the discovery and exploration of Antarctica has not been free from controversy, and argument raged back and forth for quite some time over the twin questions of who first sighted the continent and who first landed upon it. Needless to say, national pride was the fuel that fired the controversy, with the English, French and Americans being the main protagonists. However, there is now a fair measure of agreement regarding the sequence of events. It is a story of determination, courage, fortitude in the face of unbelievably appalling conditions, and more than one disaster. Altogether there have now been around 300 expeditions to the Antarctic and clearly this is not the place to discuss them all. But it would be quite wrong not to give some space to the most important and interesting events.

Do we really know who first saw Antarctica? About 650 AD a canoe commanded by a Polynesian called Ui-te-Rangiora, sailed into a place of bitter cold where the sea was covered with *pia* (a white powder) and 'things like great white rocks rose high into the sky'. Clearly these intrepid canoeists got into pack-ice waters; but did they get within sight of the Antarctic Continent? Most probably not, but in any event we shall never know. It is, however, quite likely that they were the first men to see the penguins of the Antarctic. In 1488 Bartholomew Diaz and his crew passed the southern tip of Africa and sailed on into the circumpolar ocean beyond. They were certainly the first white men to feel the cold breath of the Antarctic, but did they catch a glimpse of the continent itself? Once again, we shall never know. In the ensuing centuries other mariners venturing into these seas were, like Coleridge's Ancient Mariner, driven south by the powerful winds until 'there came both mist and snow and it grew wondrous cold . . .' So far as the record goes, however, it was Captain James Cook in the 1770s who was the first to cross the Antarctic Circle.

The cartographers of eighteenth-century Europe envisaged the then unknown southern continent as a land of green pastures. It was a French naval officer, Jean Bouvet de Lozier, in his ship the *Aigle* who, in 1738–9, discovered the island now known as Bouvet Øya and was the first to visualize Antarctica as

Portraits of Queen Alexandra and King Edward still hang in Shackleton's hut from 1914.

Adelie Penguins surround Shackleton's hut at Cape Royds.

it really was – an inaccessible continent of wind and snow girdled by pack-ice. He was also the first to give a precise description of the penguins and seals that he encountered in great numbers along the fringe of the pack-ice.

It is ironical that the most important contribution to the history of discovery of the Antarctic Continent should have been made by a man who at first doubted the existence of such a landmass. That man, as already mentioned, was the famous English seaman Captain James Cook who spent three years (1772–5) in his ships *Resolution* and *Adventure* searching for this southern continent. This voyage, consisting of three great probes to the south, marked the dawn of exploration in the Antarctic. This voyage is described by Alan Villiers in his book *Captain Cook, the Seaman's Seaman* (1967) as the greatest voyage any man has ever made.

It was on 17 January 1773 that Captain Cook first crossed the Antarctic Circle in the *Resolution*, but his way was blocked by ice and he had to turn back. In November that year he came south from New Zealand and sailed eastwards along the edge of the ice searching for an opening. Just beyond the eastern end of what we now know as the Ross Sea, he again, on 20 December, managed to cross the Antarctic Circle. On this occasion he was able to remain south of it for three days before the ice again gripped his ship and forced him to retreat. About 25 degrees further east he crossed the circle for the third and last time and reached the furthest south ever attained at that time (71°10′S) before being turned back. He did not know that on his first and third crossings he had got within about 160 kilometres of the shores of the Antarctic Continent. It was during this epic voyage that Cook discovered the South Sandwich Islands. Cook had a pretty shrewd idea of what this southern continent would be like. Take, for example, this entry in his log for 21 February 1775:

> Lands doomed by Nature to perpetual frigidness; never to feel the warmth of the sun's rays; whose horrible and savage aspect I have not words to describe. Such are the lands we have discovered; what then may we expect those to be, which lie still farther to the South?

These words penned by Cook, and perpetuated by many later writers, tend to create the impression that the Antarctic is always frightful, which indeed it is for most of the time. But there are rare occasions when this is not the case, and it is as well to set the record straight by quoting once again, from the journal of James Ross:

> Yet it can be transformed within minutes to an Eden of ethereal beauty. Quite suddenly the wind will drop, the sky will clear, the light will strengthen until mountains 300 miles away can be seen by the naked eye, and the ice will glow with colours so brilliant and be encompassed by a stillness so absolute that they have to be experienced to be believed. At such moments the Antarctic offers a pageant of beauty unequalled anywhere in the world.

At this stage in the story let us return to the question of who first sighted the Antarctic Continent, at least so far as the written record can enlighten us. There now seems little doubt that this honour must go to the Russian navigator Thaddeus Bellingshausen with his ships the *Vostock* and the *Mirnyi*, and the entry in his log for 27 January 1820 is of particular interest:

Scott's hut in its lonely setting at Cape Evans.

> At 4 a.m. we saw a grey, smoke-coloured albatross. At 7 a.m. the wind shifted, the snow momentarily ceased and we had a fleeting glimpse of the sun. We proceeded south ... At midday in Lat. 69°21'28"S, Long. 2°14'50"W we encountered icebergs which loomed up through the falling snow looking for all the world like white clouds. There was a moderate northeast wind and a heavy northwest swell at the time, and because of the snow we could see only a short distance. We had just hauled close to the wind on a course of south by east, when we observed in front of us a solid stretch of ice, running from east to west. Our course was taking us directly onto this icefield, which seemed to be covered with small hillocks. The barometer had recently fallen from 29.5 to 29, a warning of bad weather, and I therefore turned away to the northwest, hoping that in this direction we should find no ice.

The significance of this entry is that Bellingshausen's account of the icefield accurately describes that found off the coast of Princess Martha Land (a part of Queen Maud Land) which the bad visibility prevented him from seeing clearly. However, this fact, along with the known position of his ships and his subsequent references to 'an icy continent', suggests beyond reasonable doubt that the crews of these two ships were the first men to set eyes on Antarctica. By circumnavigating the continent on a course well to the south of Cook's, Bellingshausen enabled cartographers to delimit more accurately the size of the continent. Some historians have claimed that Edward Bransfield's sighting of the tip of the Antarctic Peninsula from the brig *Williams* on 1 February 1820, or alternatively the sighting in November 1820 by Nathanial Palmer in the sloop *Hero* of that part of the coast of the peninsula later named Palmer Land by Bellingshausen, should rank as the discovery of Antarctica. However, the argument becomes academic in the light of recent ice-depth soundings which have shown that most of the peninsula is an island, separated from the mainland by ice extending below sea level in a strait cutting across from the Bellingshausen Sea to the Ronne Ice Shelf. So strictly speaking, the Antarctic Peninsula is not part of the continent proper.

Although great progress had been made the protective ring of pack-ice had still not been fully breached. The breakthrough finally came in 1841 when Sir James Clark Ross with his two double-hulled ships *Erebus* and *Terror*, specially constructed for the purpose, pushed through the ice floes on 9 January to emerge in the open water inside. Having forced his way through this ring of ice, Ross sailed on southwards to enter the sea that would later bear his name, and on 11 January he sighted land. In his report he wrote: 'It rose in lofty peaks, entirely covered with perennial snow.'

He named this range on the west side of the Ross Sea the Admiralty Range, and the headland at its foot he called Cape Adare. Unable to reach the mainland, he landed instead on what he named Possession Island and took possession of the land in the name of Queen Victoria. Continuing southwards along this coast which he named Victoria Land, he came in sight, on 27 January, of the two volcanoes which he called Mounts Erebus and Terror after his ships. The mountains rose out of what was later found to be an island roughly triangular in shape, subsequently named Ross Island. Shortly after this the ships sighted what we now know as the Ross Ice Shelf, a perpendicular cliff of ice that slowly increased in height the nearer they approached to it. Although Ross did not know it at the time, the ice extended in an unbroken front for more than 800 kilometres eastwards. Ross and his men were the first to see this great barrier of ice, and they had reached further south than anyone else before. But the ice imposed a limit on further progress and they, like others who came later, ranged along it for many days before being finally forced to turn back by the formation of winter ice.

Most of the history attached to the Ross Sea area is in fact centred on Ross Island. For example, Scott's *Discovery* expedition of 1901, his *Terra Nova* expedition of 1910, Shackleton's *Nimrod* expedition of 1907 and Amundsen's 1910 expedition were all based upon it. It was across the Ross Ice Shelf that the South Pole was first reached by the Amundsen and Scott expeditions in the austral summer of 1911-12. But all of this was still a long way off in 1841. At the beginning of the present century there was still no knowledge of what lay to the south beyond the range of vision from Ross Island.

In these days of space-age technology it is easy to underrate the great achievements of these early explorers of the Antarctic. They had only great personal courage and powers of endurance, and had no power but the wind in their sails and the ocean currents to propel them through the gale-lashed seas and the ice. They had no way of knowing what they would find as they penetrated further and further into the ice, or even that they would be able to fight their way out again. We can only imagine the thoughts and feelings of Ross and his men as those snow-covered mountains came into view – the official reports give little clue to these personal aspects. In retrospect it seems almost miraculous that they and their flimsy ships survived to tell the tale. Ross's two ships, particularly well built though they were, nevertheless had a tough time. Just how tough becomes clear when we read Ross's reports, for example:

> Our ships rolled and groaned amidst the blocks of ice, over which the ocean rolled its mountainous waves, throwing huge masses one upon another then burying them deep beneath its foaming waters, the while dashing and grinding them together with fearful violence. The awful grandeur of such a scene can be neither imagined nor described.

We must turn now to another momentous development in the history of the Antarctic, the first landing on Antarctica itself, an honour which belongs to a Norwegian. On 24 January 1895 a Norwegian whaling ship, the *Antarctic*, under Captain Leonard Kristensen, hove to among the ice floes off Cape Adare and sent ashore a boatload of seven men. Some authorities, however, have suggested that the first landing took place over seventy years earlier, on Wednesday 7 February 1821. This landing was made by an American sealer, Captain John Davis of New Haven, Connecticut, in his ship the *Cecilia*. This is the entry in his log for that date:

> Commences with open Cloudy Weather and Light winds a standing for a Large Body of Land in that direction S.E. At 10 A.M. close in with it, out Boat and Sent her on Shore to look for Seal. At 11 A.M. the Boat returned but found no Seal. At noon our latitude was 64°01 South. Stood up a Large Bay, the Land high and covered intirely with snow ... I think this Southern Land to be a continent.

What this entry clearly describes is a vessel standing into Hughes Bay at the mouth of the Gerlache Strait off the Antarctic Peninsula, and landing a boat in the vicinity of Cape

The interior of Scott's hut at Cape Evans showing copies of the Illustrated London News, *a partially-stuffed Emperor Penguin and other items.*

Charles. If it is accepted, as mentioned earlier, that the Antarctic Peninsula is not part of the Antarctic Continent, then Davis was clearly not the first to land there.

The closing years of the nineteenth century saw an important change in the method of exploration of the Antarctic – the change from exploration by sea to exploration on land. The last men to explore Antarctica by sea were those of the expedition led by a Belgian, Lieutenant Adrien de Gerlache de Gomeroy in the *Belgica.* This expedition was in the field from 1897-9, and the *Belgica,* trapped by pack-ice southwest of Alexander Island, became the first exploring vessel to winter in the Antarctic, albeit unwillingly. The ship remained trapped in the ice for 347 days from February 1898, and as winter set in the men began to suffer from anaemia, paranoia and various other delusions and disorders. Their meteorological records show that during this year in the ice it was foggy on 255 days, snowed or sleeted on 271 days and was overcast on 282 days. Add to this the cold, wind, humidity, isolation and perpetual darkness, and we have a pretty graphic picture of the grim realities of life in the Antarctic winter in those early days.

The inauguration of the modern history of antarctic land exploration can be said to have begun with the British expedition of 1898-1900 led by the Norwegian Carsten E. Borchgrevink in the *Southern Cross.* He had been with Kristensen on the *Antarctic* and had landed with him at Cape Adare. Now leader of the British expedition, he landed there again on 17 February 1899. He and nine other men spent the winter there, were picked up by the *Southern Cross* on 28 January 1900 and continued south along the Ross Ice Shelf. Landing on the ice he and his party managed to advance some 26 kilometres from its edge to the record latitude of 78°50′S, thus setting the stage for

Mount Erebus looms behind Scott's hut at Cape Evans.

the penetration of the great white continent.

Up to this point the British had played the dominant role in the sea-based exploration of the Antarctic, but with the change from marine to land-based exploration it was the Norsemen with their greater experience of travel on snow in sub-zero conditions who came to the fore. There can be few people who are not aware of at least the main details of the most famous journey of all – the race to be first to the South Pole.

The two contenders were the Norwegian Roald Amundsen and the Englishman Robert Falcon Scott. On 20 October 1911 Amundsen set out from his Ross Sea base on the ice in the Bay of Whales (this no longer exists as the shores have long since calved away) with 5 men, 4 sledges and 52 husky dogs, and on 14 December they reached the South Pole; they had had a remarkably trouble-free journey. Scott established his base at the opposite side of the Ross Ice Shelf, halfway between Cape Royds and Hut Point on the southwest corner of Ross Island; he named the spot Cape Evans. The British party consisting of Scott, Henry Bowers, Edgar Evans, Lawrence Oates and Edward Wilson, set out from Cape Evans and finally reached the South Pole on 17 January 1912, only to find the Norwegian flag already flying there. Their bitter disappointment is reflected in the words that Scott wrote in his diary for that day: 'Great God, this is an awful place and terrible enough for us to have laboured to it without the reward of priority.' The three initial survivors of this tragic journey – Scott, Bowers and Wilson – starved and frozen in the interminable blizzards, died only 18 kilometres from food and safety. The hut that Scott's party built at Cape Evans remains intact to this day. A visit to this hut was one of the highlights of Eric Hosking's second Antarctic trip.

> Never [he says] have I been in a place that had such a haunted atmosphere. There was Scott's bunk with the sheepskin cover turned back as though he had just vacated it; bottles of medicine stood on the shelves; on the table was a partially stuffed Emperor Penguin and open magazines. It was as though the chaps had only recently gone out, and I could not throw off the eerie feeling that at any moment they would walk through the open door.

There have been many more expeditions since those of Amundsen and Scott. The height of activity was reached in the International Geophysical Year (IGY) of 1957-8 which recognized Antarctica as a unique natural laboratory. More than 60 research stations were set up within or very close to the Antarctic Circle. The results of the research into the geophysical and geomorphological characteristics of the continent were pooled for the benefit of mankind in general. During IGY almost 100 men managed to winter, in varying degrees of discomfort, in the very heart of Antarctica. More or less coincident with the IGY was the formation of the Scientific Committee on Antarctic Research (SCAR), and in 1959 came the signing of the Antarctic Treaty. The Treaty applies to the area south of the 60 degree South latitude, including all ice-shelves. The Treaty Powers, which include all those with an interest (including territorial claims) in the Antarctic, are currently discussing the future of the area, with particular reference to the possible exploitation of mineral resources and the harvesting of marine resources such as krill. At present the only commercial activity on the continent is tourism, which has in recent years been steadily increasing.

Whilst man could take pride in the early exploration achievements in the Antarctic, there was precious little to be proud of in the commercial activities that very soon followed these discoveries. During the 1800s the indiscriminate slaughter of the fur seals and elephant seals reduced their numbers almost to the point of extinction, and finally brought the antarctic sealing industry to a halt. Commercial whaling began at the turn of the century, with very similar results. The drastic decrease in whale stocks led to the formation of the International Whaling Commission (IWC) in 1946. Despite a somewhat unsatisfactory early record, the IWC seems to be slowly moving towards more effective conservation measures. It is to be hoped that the Antarctic Treaty Powers, despite conflicting territorial claims and other problems, will be able to reach agreement to maintain the environmental integrity of the Antarctic, and that human greed will not again hold sway in these icy waters.

The Antarctic and its Ecology

The Antarctic as it has been defined in the introduction, can usefully be divided into three main regions. These are the Antarctic Continent (Antarctica), the Maritime Antarctic and the Subantarctic. Some scientists find it necessary to go further than this and recognize eight 'provinces' within these main regions.

The uninitiated sometimes have the idea that because the Arctic and the Antarctic are both polar areas at high latitudes, albeit at opposite ends of the globe, then they must be similar in most respects. In practice there are in fact a number of very fundamental differences, and, it is of interest to compare the two polar regions. In some respects there are similarities. Both areas, for example, present severe environmental conditions for the plants and animals that live there, but conditions in the Antarctic are usually much more severe.

Whereas the Antarctic (with the exception of its oceanic islands) is a continent surrounded by ocean, the Arctic is essentially an ocean basin surrounded by mostly continental landmasses on which there is a circumpolar belt of tundra and polar deserts with North American and Eurasian elements. The ice surrounding the North Pole is basically a layer of sea-ice floating on salt water with a temperature above freezing point. In contrast, the area around the South Pole is covered by freshwater ice resting on rock; in addition the South Pole is far inland and lies at an average altitude of about 3000 metres. Even though icebergs are common to the seas around both the poles, those found in the Antarctic are far superior to the finest that the Arctic can produce. Calving ice-shelves in the Antarctic produce mighty tabular icebergs, characteristically flat-topped, as much as 160 kilometres across and many metres in height. These great icebergs sometimes seem to glow with an azure light.

When we come to compare the flora of the two polar regions we find marked differences. An almost closed cover of tundra plants, often dominated by what the botanist calls phanerogams (these are seed plants whose reproductive organs, such as cones or flowers, are clearly distinguishable), is of common occurrence around latitude 70°N, and in some instances even nearer the North Pole. Tundra vegetation occurs as far north as Peary Land on the northern tip of Greenland, reaching its limit with purple saxifrage *Saxifraga oppositifolia* at Kap Morriss Jesup at 83°39′N. In complete contrast, not a single flowering species has been recorded south of latitude 70°S. In order to find in the Antarctic a flora in any way comparable to that of the Arctic tundras, one must go northwards to the subantarctic islands. On the Antarctic Continent and the Antarctic Peninsula and its associated islands, lichens are widespread and more numerous

A massive sculptured iceberg rises from the cold waters of the Southern Ocean.

Above left *Prolific growth of the lichen* Calophaca regalis *on King George Island.*

Above right *A green carpet of moss, mainly* Calliergidum austro-stramineum, *on King George Island.*

Below *Lobster Krill* Munida gregaria *is one of the larger species of this family.*

in terms of species than are mosses. In some places the former constitute the only terrestrial vegetation. As it happens, we find among the lichens many genera (and a number of species) common to both the Antarctic and the Arctic. The fungus flora of the Antarctic Continent and Peninsula (barely a dozen species) is undoubtedly the poorest of any land area of comparable size in the world. Nevertheless, nearly all the genera, and even some of the species, are also to be found in the Arctic.

When we come to the fauna of the two polar regions we find even more striking differences. With the exception of species introduced by man, all the mammals of the Antarctic are marine. Arctic animals such as the Arctic Fox *Alopex lagopus* and Polar Bear *Ursus maritimus* (which ranges over the whole extent of the circumpolar arctic lands and sea-ice), Muskox *Ovibos moschatus*, and the various species of lemming *Lemmus* and *Dicrostonyx*, for example, have no antarctic equivalents. Lower down the scale, even the relatively species-rich freshwater habitats of the Maritime Antarctic are impoverished compared with those of corresponding arctic latitudes. Aquatic insects and fish are absent, and food chains are short. Although there are fewer species of marine mammals in the Antarctic than in the Arctic, the actual numbers are substantially larger in the Antarctic.

The contrast between the birdlife in the two areas is particularly interesting. In computing the number of different species breeding in the Arctic, it is possible to arrive at differing totals according to how one defines the Arctic. However, if the Arctic is taken to include all those areas lying north of the subpolar boreal forest, a boundary which more or less coincides with the July isotherm of 10°C, then there are approximately

158 breeding species of birds. A glance at Appendices 1 and 2 (pp. 155 and 156) shows that only 55 species and subspecies breed in the Antarctic, including those on the subantarctic islands. The obvious question is: are there any breeding species common to both the Arctic and the Antarctic? The answer to this question depends upon which taxonomic school of thought one decides to follow. Some authorities regard the Great Skua *Catharacta skua* of the northern hemisphere, and the Brown Skua *C.lonnbergi* and South Polar Skua *C.maccormicki* of the Antarctic as being simply races of one species, which then enables them to say that it is the only species in the world to breed in the high latitudes of both the northern and southern hemispheres. On the other hand many people, the writer included, prefer to consider them as separate species, and if this is so then there are no breeding species common to both the polar regions of the world.

In the southern hemisphere there is a single circumpolar ocean (the Southern Ocean) from the Antarctic Circle to about 35°S, and the only interruption is the projecting tip of South America. In view of this it is not surprising that more pelagic birds occur in the Southern Ocean than anywhere else. Their ranges tend to be defined by latitudinal rather than longitudinal limits, each species being found around the globe within certain belts of latitude. For a similar situation to exist in the northern hemisphere would require the existence of a circumpolar ocean extending southwards as far as the Mediterranean. As it is, species found north of the Arctic Circle are generally circumpolar in their distribution because they are living in a mainly unbroken expanse of ocean bounded by continents to the south. South of the Arctic Circle these continents separate the oceans and prevent the circumpolar spread of pelagic birds.

Finally, the bird life of the Arctic, by contrast with that of the Antarctic, has in the course of its evolution been obliged to adapt to the threat of such predators as the Arctic Fox and other animals, whilst as we have seen, there have never been any such predators in the Antarctic. The penguins of the Antarctic with 50 million years of evolution behind them have never, in all that time, had to adapt to the presence of any predator above water, until, that is, the recent arrival of man.

A predatory skua flying over an Adelie Penguin colony.

Massive development of tussock grasses on Macquarie Island.

Above *Wilson's Storm Petrel 'walking' on the sea.*

Overleaf *Katabatic winds blow powdered snow into the air at McMurdo Sound.*

The Antarctic Continent

For ecological purposes the Antarctic Continent includes not only the main land mass, but also the east coast of the Antarctic Peninsula south of about 64°S, Peter I Øy, Balleny and Scott Islands. Captain Robert Falcon Scott suffered from no illusions about the nature of the Antarctic Continent, which he described as 'this awful place'. Awful it certainly is for most of the time, but it is also magnificent, inspiring, and scientifically of great interest.

The Antarctic has the unenviable reputation of being the coldest, and very probably also the windiest of the world's continents. The coastal areas are generally warmer and more hospitable than the interior, but even that is not saying a great deal. The Russians have a research station lying as far into the interior as the South Pole. Lying at an altitude of 3488 metres above sea level, it holds the world record for cold at −88.3°C. Even at the height of summer its mean temperature is only −32.7°C. In coastal areas the winter temperatures are rarely lower than −50°C, and summer ground temperatures can reach 15–20°C. The continental ice-sheet is the source of cold winds which blow downslope (katabatic winds) towards and beyond the warmer coastal fringe. These winds can be very strong, and on the Adelie coast, for example, the mean annual windspeed exceeds 40 knots (20 metres per second). Not withstanding its great landmass and the enormous burden of snow and ice, the Antarctic Continent is characterized by a paucity of water in the liquid state. Even the relatively favourable snow-free coastal areas remain deserts because there is insufficient groundwater to support plant or animal life.

The area of the South Pole, lying the best part of 1600 kilometres from the sea, is among the driest on earth. The temperature always remains far below freezing point, so that the atmosphere contains no water vapour. Snow rarely falls from the sky, although it blows endlessly across the desolate and apparently limitless plain.

The area of the Antarctic Continent, including ice-shelves, is about 14 million square kilometres. Basically it is a massive ice shield with a radius of almost 2100 kilometres, and rimmed for the most part by mountain chains. From this circular shield of ice, glaciers are constantly flowing outwards between the containing mountains to form ice-shelves over the marginal seas. The huge ice sheet covering the continent is 2000–4000 metres thick, and the ice plateau of the interior is generally 2400–4300 metres in altitude. The weight of this mass of ice results in some strange anomalies. Byrd Station situated at 80°S midway along a line running from Ross Island to the Bellingshausen Sea, is at an altitude of 1584 metres, which is some 1200 metres lower than the South Pole. The thickness of the ice at Byrd is 2645 metres, so that the actual land is more than 1000 metres below sea level. This freshwater ice covering the continent represents the accumulation of snowfall over a period of something like 20,000 years, and it represents 90–95 per cent of all the glacial ice in the world. It was only in relatively recent years that it was realized that there was a continent under the Antarctic ice-sheet, and its actual area is probably only about 7 million square kilometres. Even today it is not known with certainty how much of what lies beneath the ice is continent, and how much mere chains of islands.

What is known is that the continent actually consists of two parts – usually called Greater and Lesser Antarctica – and the seam between them is marked by the Transantarctic Mountains, an escarpment more than 3000 kilometres long with a number of peaks in the 3000–4000-metre range. The highest mountain on the continent is actually the Vinson Massif in the Sentinel Range which rises to 5140 metres. The enormous ice dome of the main part of the continent conceals a great mountain range of more than 3000 metres elevation, and more extensive than the European Alps. A few of the higher mountain peaks and ranges of the interior project above the ice as nunataks, that is to say areas that have never been completely inundated by ice.

The overwhelming majority of the Antarctic Continent consists of a sterile ice-desert surrounded by a coastal fringe whose few snow-free areas are desert. Altogether, only some 2–3 per cent of the land surface is ice-free. A Soviet geographer has calculated that of the total 30,000-kilometre coastline of the continent, 9 per cent is still formed by ice-cliffs; backed by floating shelf-ice (45.5 per cent); active glacier (9.5 per cent); or static ice resting on land. Less than one per cent of the exposed ground on the continent has plant cover of any kind.

One of the most intriguing phenomena are the ice-free areas known as dry valleys. Extensive dry valleys or oases are found in Victoria Land, Knox Coast, Queen Mary Coast, Pravda Coast and Queen Maud Coast. These dry valleys lie below what we might call, for lack of a more suitable term, the 'snowline', which is the point at which precipitation equals sublimation. That is to say that nearly all of the snow that falls evaporates to vapour, and only a small proportion actually melts.

One of the largest of these ice-free areas is known as the McMurdo Sound oasis, parts of which are the driest on this planet. Its mean annual temperature is about –20°C. The McMurdo oasis includes a number of dry valleys with enclosed drainage basins. The lowest part of each of these valleys is occupied by a saline lake, other examples of which can also be seen in nearby Victoria Land. A characteristic feature of these unique lakes is a deep, warm-water layer. The lake surfaces are irregular and etched with meltwater patterns. Although permanently frozen, they melt in summer for a few metres around their margins. The flora of these lakes includes algae and diatoms. Highly saline soils are characteristic of all the dry valleys.

These dry valleys represent a unique biological system. The prevailing combination of environmental factors found there – for example, dessication (lack of available water), limited maximum temperature, frequent freeze-thaw cycles, very short growth periods, and so on – means that the life found there consists almost exclusively of micro-organisms. The bulk of the microbial life is located in stream beds or ponds fed by alpine glaciers, or in the poor soils in the immediate vicinity of the glaciers.

Even in the ice-free areas of Antarctica the soils remain primitive in character, and mature organic soils are absent even in the most favourable sites. Deposits of a peat-like character up to several centimetres in depth have, however, been found as far south as McMurdo Sound (77°30′S). Where penguins concentrate to breed, richly ornithogenic soils occur. These are soils greatly enriched by the continuous droppings of the penguins. Given the combination of a harsh climate and extremely poor soils, it is hardly surprising that no flowering plants occur. The most prominent plants are algae, lichens, and to a much lesser extent mosses. All these groups are represented only 260 kilometres from the South Pole at altitudes up to almost 2000 metres on exposed rocks among the peaks of the Horlick and Queen Maud Mountains.

The natural vertebrate animal life of the Antarctic Continent consists of seabirds and seals. They are all animals of the coasts or the adjacent pack-ice environment. The sea-ice is the year-round habitat of the Weddell Seal *Leptonychotes weddelli*, Crabeater Seal *Lobodon carcinophaga* and Ross Seal *Omatophoca rossi*, but only Weddell Seals are normally to be found ashore or close to land. The Leopard Seal *Hydrurga leptonyx* is also closely associated with the sea-ice environment. Ten species of bird breed or have bred on the main mass of the Antarctic Continent, these being the Emperor Penguin *Aptenodytes forsteri*, Adelie Penguin *Pygoscelis adeliae*, Southern Giant Petrel *Macronectes giganteus*, Southern Fulmar *Fulmarus glacialoides*, Antarctic Petrel *Thalassoica antarctica*, Cape Pigeon *Daption capense*, Snow Petrel *Pagodroma nivea*, Wilson's Storm Petrel *Oceanites oceanicus*, South Polar Skua *Catharacta maccormicki* and Antarctic Tern *Sterna vittata*. The Antarctic Prion *Pachyptila desolata* is now thought to be extinct as a breeding species on the continent. The Chinstrap Penguin *Pygoscelis antarctica* must be included as a breeding bird of the continental region by virtue of colonies on at least Balleny Island and Peter I Øy.

Only three of these birds, the Antarctic and Snow Petrels, and the South Polar Skua, are known to nest deep in the

Above *The Antarctic Petrel is one of the limited number of species that breeds on the Antarctic Continent.*

Right *A Macaroni Penguin showing the extaordinary development of yellow feathers on the head.*

interior of the continent. All others, like the penguins, nest around the coasts. The primary species of the pack-ice environment are the Emperor Penguin (which actually breeds on the ice) and Adelie Penguin, and the Antarctic and Snow Petrels. Neither of the petrels normally go beyond easy flying distance of the ice. Open water in the form of leads and cracks in the ice occur as a result of pressures and tensions within the ice-sheet, or as a result of tidal movements where the ice joins the land. They are important in the ecology of the birds and seals as they provide feeding areas for the former, and allow the latter to breathe and to haul out of the water to lie on the ice.

Pack-ice surrounds the Antarctic Continent northward to and, in some cases beyond latitude 60°S. In winter and early spring the pack-ice reaches its northern limit, an average distance of some 800 kilometres from the coast. The sea area thus enclosed within the limits of the ice is some 19 million square kilometres, or very nearly half as much again as the area of the continent itself. Every winter the sea freezes to a depth of up to 3 metres to a distance of up to 200 kilometres offshore. Much of this ice is what is known technically as *fast ice*, because it is anchored to the land, in which state it may remain throughout the winter. This situation prevails all around the continent, except in areas known as *polynyas* where powerful offshore winds or currents keep the sea clear of ice even in winter. In contrast to the rough and stormy seas and grey skies beyond, the pack-ice world is calm, white and glaring, with clouds above as white as the snow and ice below, and open leads of water seeming ultramarine in colour. Most of the Ross Sea is covered by a shelf of ice greater in area than France. The seaward edge of this shelf is a gleaming ice cliff up to 60 metres high, and stretching for more than 800 kilometres.

The Maritime Antarctic

The Maritime Antarctic, the second of our three antarctic regions, covers a total area of about 40,000 square kilometres. Even though the climatic conditions are less severe than those prevailing over most of the Antarctic Continent, approximately 90 per cent of the maritime region consists of permanent snow and ice-fields.

Seen from the viewpoint of an orbiting satellite, the Antarctic Continent is comma-shaped with a long tail (the Antarctic Peninsula) pointing towards the southern tip of South America. The mountainous peninsula extends for 1600 kilometres in a narrow sinuous curve varying from 45 to 260 kilometres wide. Its plateau is more than 2000 metres high in the south, but drops gradually towards the northern extremity. The highest peak is Mount Jackson at 72°S which rises to nearly 3500 metres. The climate of the peninsula is the coldest and driest in the Maritime Antarctic.

The Maritime Antarctic comprises the west coast of the Antarctic Peninsula and the offshore islands to about 70°S, the South Sandwich, South Orkney and South Shetland Islands, and also Bouvet Øya. The region has a history of volcanic activity and Deception Island in the South Shetlands, several of the South Sandwich Islands, and Bouvet Øya are still volcanically active. In these localized active areas, escaping gases and mineral solutions warm the soil thus allowing colonization by mosses and microscopic forms of animal life. As may be expected, this region has a cold maritime climate and in summer the mean monthly temperatures exceed 0°C for up to four months in the north, and for one to two months in the south. Winter temperatures rarely go below −10 to −15°C. The coldest of the maritime islands are Adelaide Island, the Palmer Archipelago, and the other islands that lie off the west coast of the Antarctic Peninsula. Although most of the islands in this region are close to the Antarctic Peninsula, they are surrounded by open water in summer, but are linked to the mainland by pack or fast-ice in winter.

The South Shetland Islands form an extensive archipelago running parallel to the west coast of the Antarctic Peninsula. They consist of the exposed peaks of an arc of submerged mountains, forming three large and almost a dozen smaller islands. Most are covered by permanent snow and ice-fields, and the soils have rarely developed beyond a primitive organic stage. The South Orkney Islands comprise four main ice-covered islands and a number of satellite islets. The largest of the group is Coronation Island whose dominant feature is Mount Nivea (1265 metres). Most of the headlands and small islands off the south coast are ice-free in summer. On Laurie Island an ice-free area is the site of Orcadas, the oldest-established human settlement in the Antarctic, founded by the Scottish National Expedition in 1903. Low-lying Signy Island is the most biologically important island of the group and has only a small remnant ice-cap. The eleven islands of the South Sandwich group are spread along a 350-kilometre crescent. Only Cook, Montagu and Vindication Islands in the group seem to be volcanically inactive and cold. The entire group is much less hospitable than the South Orkney or South Shetland Islands, and only a very small proportion of the total land area is ice-free. Finally, Bouvet Øya, about 54 square kilometres in extent, rises steeply from the sea to a flat ice-covered dome 935 metres high. The island is almost completely covered with

Above *A pair of King Cormorants at their nest with Rockhopper Penguins behind them.*

Right *A Gentoo Penguin with two young. This species breeds from as far north as the Falkland Islands, south to the Antarctic Peninsula.*

glaciers and only small areas of bare rock are exposed. It is biologically barren despite the presence of some breeding penguins and seals.

The maritime islands have a slightly longer growing season than the continental islands, and there is usually more summer precipitation in the form of sleet or wet snow, so the islands are greener in appearance. From about late October to April plants and animals of the ice-free areas and freshwater habitats are able to grow actively. The lakes thaw out and the water warms to the point where there is often a rich algal growth. The soils, too, show an improvement in quality compared to those of the Antarctic Continent. Brown soils, not unlike those found in the Arctic, occur here and there together with moss-peat to depths of up to two metres as far as about 65°S. On Elephant Island in the South Shetland Islands these peat deposits may attain depths of up to three or four metres in particularly favourable locations. The organic soils are able to retain moisture throughout the summer.

Given the relatively more favourable environmental conditions compared with the Antarctic Continent, it is not surprising that the flora of the Maritime Antarctic is somewhat more diverse with a rather greater variety of species of mosses and liverworts, for example. Here, too, flowering plants make their appearance for the first time, but the geographical isolation of the islands and the diurnal freeze-thaw cycle, allows only two species to occur. These are *Deschampsia antarctica* and *Colobanthus quitensis* which extend to 68°S in Marguerite Bay on the west coast of the Antarctic Peninsula. Both these flowering plants occur on well-drained, north-facing slopes in the coastal zone, often below cliffs with nesting birds. They grow mostly as isolated tufts or cushions, usually in association with mosses and other

lowly plants, but on rare occasions may form a closed turf.

The development of vegetation on the maritime islands is very much dependent on the availability of free water, and is best developed in areas where there is considerable summer snow-melt and precipitation. The most extensive stands of closed vegetation have developed on the more stable areas such as rocky ridges and headlands, and on scree slopes. In the South Orkney Islands, for example, raised beaches at heights of up to 30 metres are found, and it is on these that the most extensive stands of closed vegetation are found. Generally speaking, the islands have a rather barren appearance and moss and lichen communities are sparse. Wet habitats are dominated by mosses, and dry rock surfaces and 'fellfield' habitats by other mosses and also lichens. The latter tend to predominate in exposed situations and inland. Snow algae and species of the larger fungi are frequent in summer. On four of the South Sandwich Islands where volcanic fumaroles warm the ground, unique zoned communities of bryophytes (mosses and liverworts) are found containing species not occurring elsewhere in the Antarctic.

One of the features of the Antarctic Continent, the Maritime Antarctic, and to a lesser extent the Subantarctic, are the extensive green sheets formed by a green foliose alga *Prasiola crispa*. These occur on the nitrogen-rich muddy soil around (but not in) penguin rookeries, and also seal rookeries, giant petrel colonies and below cliff-breeding petrels.

The Maritime Antarctic has an abundant marine-bird and mammal fauna comprising species which rely on the sea for virtually all their food, but most come to land to breed. There is also a substantial invertebrate fauna which, unlike that of the Antarctic Continent, includes representatives of the higher insects in the form of two species of chironomid midges. The South Sandwich Islands can boast at least a dozen species of land arthropods that are not known to occur south of latitude 60°S. Some of the ice-free coastal plains of these islands also support huge penguin rookeries. Overall, the South Shetland Islands emerge as the richest biologically and the most favoured ecologically of those islands lying south of latitude 60°S.

The relative biological richness of the South Shetland Islands manifests itself in several ways. It is here (and on the adjacent peninsula coast), for example, that the two species of Diptera mentioned above are found. The arthropod and nematode faunas both exhibit clear gradients of decreasing diversity as one moves southwards and eastwards from these islands. Another indication is found in the bryophyte flora, many species of which have a higher reproductive success in the South Shetland Islands than is the case in the South Orkney Islands.

A total of 17 species of birds breed in the Maritime Antarctic, including all those that also breed on the Antarctic Continent with the exception of the Antarctic Petrel. The Emperor Penguin qualifies as a breeding bird of this region solely because of a small colony on the Dion Islands at 67°52′S. This is the only known breeding site for this species on the west side of the Antarctic Peninsula. The Maritime Antarctic has eight species that do not breed on the continent, these being the Gentoo Penguin *Pygoscelis papua*, Macaroni Penguin *Eudyptes chrysolophus*, Antarctic Prion, Black-bellied Storm Petrel *Fregetta tropica*, Antarctic Cormorant *Phalacrocorax atriceps*, Dominican Gull *Larus dominicanus*, Brown Skua and the Wattled Sheathbill *Chionis alba*. Both the Brown Skua and the Wattled Sheathbill can be regarded as vertebrate carnivores since they scavenge, take the eggs and occasionally kill the young of other birds. Of

the four species of breeding penguins, it is the Chinstrap that is the characteristic penguin of the Maritime Antarctic, and both its numbers and range have been increasing in recent years, and its current population is numbered in millions. The various species of petrel breed in thousands on the steep cliffs, and their droppings encourage the growth of algae and lichens.

Several species of seal occur, and perhaps the most interesting of these is the Leopard Seal *Hydrurga leptonyx*. Where, for example, do most Leopard Seals breed? The general concensus among experts is that it is a widely dispersed breeder in areas of broken ice, and its main breeding area is in the southern part of the Maritime Antarctic; but nobody knows for certain. The Southern Elephant Seal *Mirounga leonina* has its main breeding colonies in the Subantarctic, but large numbers haul out in summer on beaches in the South Orkney and South Shetland Islands where small satellite breeding groups have been found. It also breeds sparsely in the South Sandwich Islands and a few, probably less than 100 in any one season, on Bouvet Øya.

The Weddell Seal may not be a numerous breeding species in this region because its distribution is probably influenced mainly by transient ice conditions. Certainly, in the South Orkney Islands, breeding seems to be a rare event with pupping occurring only where ice conditions are favourable. Not too much seems to be known about the breeding distribution of the Crabeater Seal in the Maritime Antarctic, but pupping is known to take place in heavy pack-ice areas. The Antarctic Fur Seal *Arctocephalus gazella* is the most southerly representative of this genus, reaching the South Shetland Islands where it breeds. It breeds also in the South Orkney Islands, South Sandwich Islands and on Bouvet Øya, but is mainly a species of the Subantarctic.

Above *The Weddell Seal inhabits the inshore fast-ice zone where it survives the winter in the slightly warmer water below the ice.*

Left *This green alga,* Prasiola crispa, *often forms extensive sheets, especially on the soil around penguin colonies.*

The Subantarctic

Situated as they are on or reasonably close to the Antarctic Convergence, the subantarctic islands of South Georgia, Marion and Prince Edward Islands, Îles Crozet, Îles Kerguelen, MacDonald and Heard Islands, and Macquarie Island, are the warmest of the antarctic islands and are free of sea-ice throughout the year, except for small areas which may form in bays and fiords, as at South Georgia. The islands have a cool, cloudy and windy oceanic climate with consequently greater precipitation (much of which falls as rain) than the two more southerly regions that we have already discussed. The mean monthly temperature is above freezing point for at least half the year. With the exception of Îles Crozet and Macquarie, the islands are ice-capped, but at sea level the winters are relatively mild and the summers long and cool. All the islands are characterized by predominantly herbaceous vegetation in lowland areas forming locally extensive closed communities. These grade to open fellfield vegetation of scattered grasses, cushion plants, mosses and lichens on more windswept terrain.

The map shows a long, sickle-shaped string of islands (the Scotia Arc) stretching from the tip of South America towards the Antarctic. This string of islands, including the South Sandwich, South Orkney and South Shetland Islands, forms a link between the South American Andes and the great mountains of the Antarctic Continent. South Georgia, along with the rest of these islands, is composed of the exposed peaks of a great submarine mountain range. It is the largest and highest of the islands on the Scotia Arc.

The island is mountainous and huge towering peaks mantled with white sweep straight up from the sea. The peaks of its central massif rise in a spectacular ridge to more than 2700 metres. The surrounding sea is, at best, never still, and at times can become unbelievably ferocious. For most of the time the mountains are shrouded in cloud, but there are days when anticyclonic weather conditions bring a clarity to the scene, and on such occasions South Georgia, seen from afar, is a breathtaking sight not easily forgotten. Obviously the weather is very often grim, and daily rainfall in summer may exceed 100 millimetres. In winter it is cold, but not excessively so, and sheltered bays may freeze over for periods of a few weeks. The weather is not quite as bad as implied some years ago in a publication issued for the guidance of mariners, which contains the jaundiced statement that 'the climate of South Georgia is uniformly dismal'.

Most of South Georgia is covered by extensive permanent ice with many large glaciers flowing down to reach the sea at the head of fiords. Despite this the island has some deep organic soils with a clear profile structure. Parts of the island have been free of glacial ice for at least 6000 years, and some peat deposits are believed to date back this far. Much of the coast is formed of high sea cliffs, but the north side of the island is indented with numerous deep bays, fiords and glacial valleys. In terms of diversity of plant and animal life, and the extent of closed vegetation, South Georgia is more or less intermediate between other subantarctic islands and the Maritime Antarctic region.

There has been a permanent human population on the island since 1904, reaching a peak in the heyday of the sealing and whaling era. The last whaling station was closed in 1965. Not surprisingly, man has had a considerable influence on the island's ecology. The whalers introduced reindeer and various other animals, but only the former survive. They have increased

Macquarie Island – Southern Elephant Seals lie in the tussock grass, while behind them stand the vats in which their carcasses were boiled for oil in the days of the sealing industry.

The skeleton of a Blue Whale, the largest living mammal in the world, on King George Island.

to around 2500 in number over the past 60 years and have had a serious local effect on the vegetation in three areas of the island. Overgrazing and trampling has led to the eradication of many macro-lichens and bryophyte banks. Some of the overgrazed plant communities have been replaced by stands of the alien grass *Poa annua*. More than 50 species of alien plants are known from the island, particularly around the sites of the former whaling stations. Some of these alien plants have become naturalized and compete aggressively with the native vegetation. Unfortunately the Brown Rat *Rattus norvegicus* reached the island and is widespread in lowland areas. It takes eggs and young from the nests of burrowing petrels such as the Black-bellied Storm Petrel, Wilson's Storm Petrel and the two diving petrel *Pelecanoides* species, and has decimated the populations of these species.

Marion and Prince Edward Islands are of volcanic origin dating from about half a million years ago. They lie 220 kilometres north of the Antarctic Convergence. The larger of the two, Marion Island, has an area of 290 square kilometres, while Prince Edward Island is only 44 square kilometres in extent. Both are situated on a submarine platform, and both have a prominent escarpment on the west, below which is a level coastal plain. There is little seasonal variation in their typical subantarctic climate, characterized by frequent gales, much cloud, and abundant rain and snow averaging 2500 millimetres each year. The islands are situated in the Roaring Forties and experience strong, almost continuous winds which average 32 kilometres per hour.

Marion Island consists of three distinct regions: a 1000-metre high central plateau much of which is snow-covered all year; the slopes running down to the coast and forming the greater part of the island; and the flat marshy coastal plain barely 50 metres above sea level. The island also has a number of small lakes, and its dome-shaped profile is interrupted by some 130 conical hills up to 200 metres high which mark the sites of former volcanic eruptions. Prince Edward Island has a more vertical relief. A dozen or more alien weed species are present on Marion Island, but only one on Prince Edward Island. Introduced mice and cats are also present on Marion Island where the population of the latter is now in excess of 2000, and may kill as many as 600,000 small petrels annually.

The Îles Crozet, an archipelago of five volcanic islands with a total land surface of about 500 square kilometres, are situated 400 kilometres north of the Antarctic Convergence. Climatically they are similar to Marion and Prince Edward Islands, but may be marginally milder. The western part of the group comprises a handful of small islands and reefs and are virtually inaccessible. One of these, Île aux Cochons, is a cloud-capped volcanic cone about 800 metres high, with high sea cliffs on the west. The two eastern islands of the group are Île de la Possession and Île de l'Est, of which the former is the largest in the archipelago and has a high interior snow-capped chain of mountains, dominated by the 934 metre-high Pic du Mascarin. Deeply eroded valleys lead from the interior to sandy beaches on the north and east coasts. The smaller Île de l'Est is a high and deeply dissected island dominated by several crests and ridges, the two highest peaks of which are 1050 metres and 1012 metres. There are five valleys the floors of which have streams and lush vegetation. Although these valleys terminate in beaches, the majority of the coastline consists of steep cliffs. Introduced rabbits have grazed extensive areas of these two islands, and are also present on Île aux Cochons.

The Îles Kerguelen, situated right on the Antarctic Convergence, comprise a compact archipelago of one large island and about 300 smaller islets of early volcanic origin with a total land area of more than 7000 square kilometres. Temperatures vary little from season to season, and precipitation is high and frequent and falls on 250–300 days in the year, mostly in the form of rain. It averages over 1100 millimetres in the south-east, and twice that much in the mountains in the west.

Île Kerguelen is the largest island (6000 square kilometres) and has a sharp high relief with extensive 200–800-metre-high plateaus, broken up by deep glacial valleys and lakes. Several snow-capped mountain peaks are more than 1000 metres high, and the twin peaks of Mount Ross rise to 1850 metres near the south coast. Only the eastern end of the island has extensive flat lowlands. A profusion of lakes and ponds of various types are to be found on the island. In the west, at Calotte Glaciaire Cook, can be seen the approximately 600-square-kilometre remains of an ice-sheet that formerly covered the entire island. So deeply dissected is the coastline that no part of the island is more than 20 kilometres from the sea. Around the coast, cliffs up to 600 metres high alternate with long beaches of sand and shingle.

Unfortunately, introduced rabbits and other mammals (including reindeer and sheep) have ravaged the native vegetation on Île Kerguelen and significantly changed the structure of the plant communities. Erosion induced by the grazing activities of these introduced animals has reduced the amount of habitat suitable for burrow-nesting petrels, whose eggs and young are also preyed upon by introduced Black Rats *Rattus rattus*.

Heard and Macdonald Islands, both of which lie on the same submarine plateau as Îles Kerguelen, are about 180 kilometres south of the Antarctic Convergence. Ninety per cent of Heard Island is covered by glacial ice up to 150 metres deep in places. Over half of the coastline is bordered by ice cliffs. The central part of the island is roughly circular in outline and dominated by the dome-shaped bulk of 2400-metre-high Big Ben, which in turn culminates in the intermittently active volcano of Mawson Peak 2745 metres high. Glacial moraines provide soil for lowland vegetation and thus nesting sites for birds.

The three islands comprising the Macdonald group are collectively smaller than Heard Island, with Macdonald Island itself being the largest of the three. The northern half of the island is a sloping plateau with a maximum elevation of 120 metres, and the southern half is a steep-sided hill about 230 metres high. Macdonald Island was landed on for the first time as recently as January 1971. Luxuriant growths of tussock grass *Poa cookii* occur on the eastern slopes of the 230-metre-high hill in the south, and on the plateau at the northern end. There are numerous freshwater tarns on the plateau, and much of the highest part is covered by *Azorella selago*. Although hundreds of Southern Giant Petrels nest in the plateau grass, and the unvegetated section is honeycombed with burrows believed to be those of the South Georgian Diving Petrel *Pelecanoides georgicus*, the most abundant bird is the Macaroni Penguin which occurs in huge numbers in many areas and may be heard and smelt from as much as 8–10 kilometres offshore! In view of the relatively inaccessible nature of the islands it is hardly surprising that no alien plants or animals have become established there, unlike all other islands in the Subantarctic.

The last island in this section of the Antarctic is Macquarie Island which has an area of 112 square kilometres, and whose rocks are mainly igneous and alkaline. No permanent snow or ice exists on the island at the present time, although glaciers once covered the entire area, gouging out small lakes and rounding the highest points. The island consists of a central

This whaling station on Deception Island in the South Shetlands was destroyed by volcanic activity in 1968.

Fin Whales break the surface of the Southern Ocean.

Southern Giant Petrel incubating. Note how large the bill is in relation to the small head.

plateau 250–300 metres in elevation, from which steep slopes with cascading streams descend to narrow raised coastal terraces as much as 800 metres wide on the south side of the island. The highest peak, Mount Hamilton, is a mere 433 metres above sea level. Seen from far out at sea, the island appears as a long, irregular, green mountain range.

The cold temperate climate of Macquarie Island is marked by heavy cloud cover most of the year, misty rain averaging about 1000 millimetres annually and distributed throughout the year, a relative humidity averaging 92 per cent, and a wind velocity in which gusts of 100 kilometres per hour are frequent all year. The one bright spot in this dreary weather pattern is that from November to March the sun may shine for more than two hours per day!

The familiar pattern of destruction by introduced animals is repeated here also. Rabbits have ravaged much of the lush native tussock grass and herbfield vegetation, and cats, rats and Wekas (a New Zealand bird) are important predators on the eggs and young of many of the breeding birds.

The half-dozen islands and island groups in our subantarctic area have a very much richer native flora than either the Antarctic Continent or the Maritime Antarctic. If we take approximate figures the region as a whole has some 70 species of vascular plants; 250 mosses, 150 liverworts and 300 lichens. Overall, South Georgia emerges top of the league with almost 450 species in these four groups, followed by Îles Kerguelen with nearly 280 species, and Îles Crozet with about 238 species. Heard and Macdonald Islands have the poorest flora in the Subantarctic. The only true shrub found in the Subantarctic, *Coprosma pumila*, is generally sparse and found only on Îles Kerguelen and Macquarie Island. On all the islands the general rule is that with increasing altitude or exposure the vegetation shows a progressive reduction in both size and vigour. There are many similarities in the vegetation of the islands, the general pattern being a coastal zone of luxuriant tussock grasses (which may reach two metres in height on South Georgia), followed by herbfield communities, then fellfield, and finally more or less bare ground with or without snow cover. Marsh, bog and pond habitats also occur frequently.

On South Georgia the tussock grass *Poa flabellata* is confined to coastal areas but may extend as high as 225 metres on some hillsides. It is particularly lush where there is nitrogen enrichment from seabirds and seals. Wet level ground between tussocks and on raised beaches is locally dominated by the grass *Deschampsia antarctica*. Another grass, *Festuca contracta*, reaches the southernmost extremity of its geographical range on South Georgia, but only develops into extensive closed grassland on sheltered slopes in that part of the island that has been deglacierized the longest. On drier more exposed slopes it forms a mixed grassland with the shrub-burnet *Acaena magellanica* together with many bryophytes and lichens. On wet valley floors and in rock basins extensive mires occur dominated by a rush *Rostkovia magellanica* together with mosses and liverworts. The dry, windswept areas that receive little or no winter snow-cover support a very open fellfield vegetation in which low cushion and turf-forming mosses and lichens are the dominant species.

On Marion and Prince Edward Islands we again find tussock grassland, dominated this time by *Poa cookii*, on well-drained coastal and inland slopes that are exposed to the effects of sea-spray and fertilized by seabirds. Better protected, well-drained inland slopes have a mixed herbfield vegetation of ferns, grasses, sedges, dwarf shrubs and cushion plants. On warm north-facing slopes the dominant species is the fern *Blechnum pennamarina* forming a fernbrake community, and protected eastern slopes may have as the dominant plant the burnet *Acaena adscendens*. Poorly drained valleys have much swampy ground carpeted with mosses, sedges and rushes. The fellfield communities above 300 metres consist almost entirely of mosses and lichens.

The vegetation of the three largest of the Îles Crozet is relatively uniform, and coastal areas have the usual luxuriant tussock grassland (*Poa cookii*). Areas of undisturbed moorland have a plant community in which Kerguelen Cabbage *Pringlea antiscorbutica*, clumps of *Acaena*, and cushions of Azorella *Azorella selago* are prominent. On the higher slopes above 300 metres only a sparse cover of mosses and lichens is to be found. For various reasons, the vegetation of the Îles Kerguelen is rather difficult to describe succinctly. Four major species – Kerguelen Cabbage, *Poa cookii*, cushions of Azorella, and the mat-like burnet *Acaena magellanica* – occur from sea level to more than 500 metres elevation in extensive herbfield communities which, however, may vary considerably in terms of species composition and relative abundance. Undisturbed tussock grassland still occurs on some of the smaller islands, and on the main island in wet upland areas around the 500-metre elevation. Two grasses, *Deschampsia antarctica* and *Agrostis magellanica*, are found along stream banks and to a lesser extent in bogs. The high altitude areas have the usual fellfield vegetation or are devoid of plant cover entirely. Heard and Macdonald Islands have relatively few plant species. Kerguelen Cabbage, *Poa cookii* and cushions of Azorella occupy valleys and old moraines up to 200 metres or so. Mosses and lichens occur on bare ground from 200 metres up to the snowline.

Macquarie Island is extensively vegetated and five main communities can be recognized – tussock grassland, herbfield, fen, bog and fellfield. The latter in fact covers something like half the island (including most of the plateau) in areas exposed to high winds. The dominant plants are cushions of Azorella and the moss *Rhacomitrum crispulum*. Vegetation cover in this habitat averages about 45 per cent, but may be as low as one per cent in extreme situations. Tussock grassland, dominated by *Poa foliosa* with Macquarie Island Cabbage *Stilbocarpa polaris* as a co-dominant species, is common on well-drained coastal terraces, on slopes up to about 300 metres, and in sheltered upland areas. From near sea level to around 335 metres, on flats and slopes where winds are moderate and the water table high, may be found lush herbfields dominated by rosettes of *Pleurophyllum hookeri* (one of the Compositae or daisy family), sometimes with Macquarie Island Cabbage as co-dominant. Alkaline fens of sedges and rushes (dominant species *Juncus scheuchzerioides*) occupy flats locally in valley bottoms and on beach terraces where the water table is at the surface. Considered by some to be the most interesting habitat on Macquarie is the extensive acid peat bog on a raised coastal terrace at the north end of the island. Here, on a deep bed of peat, is a soggy carpet of mosses and sedges through which are scattered the blue-grey rosettes of *Pleurophyllum hookeri*.

The so-called penguin-tussock *Poa hamiltoni* is very locally distributed on Macquarie Island. It grows lushly in association with rookeries of Macaroni Penguins and Rockhopper Penguins *Eudyptes crestatus*, at sites where the guano-impregnated waters trickle out of the nesting zones.

The fauna of the Subantarctic is, as we might expect, much more diverse than anywhere else in the Antarctic. The invertebrate fauna includes many higher insects, and also

Breeding area of King Penguins on South Georgia, showing a crèche with adults. Note the size of the grass tussocks in the background. The breeding population of this species on South Georgia has greatly increased during the last 30 years.

spiders and snails. On Macquarie Island alone, for example, over 119 species of arthropods have been recorded, of which about 100 or so are associated with plants or plant litter. Absorbing though these smaller forms of life may be to the scientist, to the layman who may happen to find himself in the Subantarctic, it is the seals and seabirds that dominate the scene.

Although small numbers of Leopard and Weddell Seals breed in this region, as for example at South Georgia, the most prominent species is the Southern Elephant Seal which has its largest world population at South Georgia where its numbers are increasing rapidly. Other important populations of this species are found at Macquarie Island and Iles Kerguelen. The main populations of the Antarctic Fur Seal are also centred on South Georgia, with smaller subsidiary populations on other islands of the Scotia Arc north of 65°S. The Antarctic Fur Seal has, probably within recent years, begun to breed in small numbers at Marion Island along with the Kerguelen Fur Seal *Arctocephalus tropicalis*, the fur seal of the islands north of the Antarctic Convergence, but which has a population of around 7000 at Marion Island.

A glance at Appendix 2 shows that 50 species and subspecies of birds (mostly seabirds) breed in the Subantarctic, many of them in immense numbers. Penguins, albatrosses and petrels are conspicuous members of the Subantarctic ecosystem, and calculations which have been made indicate that approximately 65 per cent of the total penguin populations of the Antarctic as a whole is found in the subantarctic region. Other calculations suggest that in this region the penguins, albatrosses, petrels (including the diving petrels) and cormorants, together consume some 21 million tonnes of food each year. These are obviously somewhat crude calculations, but they do serve to indicate the importance of seabirds in the ecology of the Subantarctic. The huge breeding concentrations of seabirds, especially penguins and petrels, have a strong influence on the vegetation in enriching the soil and causing extensive erosion, sometimes down to bedrock.

The King Penguin *Aptenodytes patagonicus* is very much a bird of the Subantarctic. The biggest breeding concentrations are at Îles Crozet, and Marion and Prince Edward Islands. This penguin has shown a considerable population increase in the past few decades, as for example at South Georgia and it has been suggested that this rise in numbers and those of the fur seal may be connected with the increased availability of food (mainly cephalopods and fish), which in turn may have been caused by a higher krill production due to the reduction in stocks of baleen whales.

From the point of view of seabird populations, the Îles Crozet are important with 32 breeding species, while Îles Kerguelen has 28 and Marion and Prince Edward Islands 27 species. On the latter islands 10 of these breeding seabirds are small petrels which breed underground and are mainly nocturnal. The total penguin population of the Marion and Prince Edward Islands probably exceeds 1.6 million with Macaroni and King Penguins being the most abundant. The world's second largest breeding population of King Penguins is also in these islands.

Head of a male Southern Elephant Seal showing the large inflatable nostrils which hang down over the mouth and form a resonating chamber. This species has extended its range southwards in recent years.

King Penguins walking in file on the stony shore of Macquarie Island.

Penguins

Seven species of penguin, all well adapted to the rigours of their environment, breed south of the Antarctic Convergence, while an eighth species, the Royal Penguin *Eudyptes schlegeli*, nests only on Macquarie Island. Among the factors contributing to the success of these birds are the long summer days, the great abundance of food, and a relatively low level of predation. From the studies that have been made it is clear that the breeding colonies are sited so as to make the best of the cold climate, but at the same time they are distributed in such a way that they are able to take full advantage of the food resources. Almost every island has more than one resident species (see Appendices 1 and 2). Since they are so obviously successful it follows that penguins must be well adapted to the harsh environment in which they live, and this is indeed the case. To start with they are superbly well insulated. Not only do they have a thick layer of insulating fat, but the feathers are more tightly packed than those of other birds, and have tips which overlap like tiles on a roof. Between the feathers and the layer of fat is a layer of dense, woolly underdown.

The accounts written by early explorers in the Antarctic nearly all contain references to these birds and today, being vivacious, inquisitive and very photogenic, they are a source of entertainment to the new generation of tourists. On 11 January 1841 when James Clark Ross was thwarted in his attempt to land on the mainland at Cape Adare, he landed instead on what he named Possession Island and found that

> inconceivable myriads of penguins completely covered the whole surface of the island, along the ledges of the precipices, and even to the summits of the hills, attacking us vigorously as we waded through their ranks, and pecking at us with their sharp beaks, disputing possession.

Edward Wilson visited Cape Adare in January 1902 and wrote in his journal that the Adelie Penguin rookery there contained literally millions of birds.

For many of the early visitors the penguins were a source of amusement just as they are today, but for some they meant the difference between life and death. This was the case with Otto Nordenskjold and his companions of the 1901-4 Swedish expedition in 1903 when faced with their second (and this time unexpected) winter on Snow Hill Island off the east coast of the Antarctic Peninsula. If it had not been for the seals and penguins they would have died of starvation. They killed only 30 seals, but more than 400 penguins. No visitor to the Antarctic can fail to be amused by the appearance and antics of the penguins. M. Racovitza, a member of the 1897-9 Belgian

Right *Adelie Penguins on Torgenson Island.*

Overleaf *A nesting colony of Adelie Penguins.*

Below *Adelie Penguins at Hope Bay. This is the most southerly breeding penguin of the three Antarctic species in the genus* Pygoscelis. *It has a circumpolar distribution on the Antarctic Continent.*

These pictures of the Adelie Penguin show three stages in the feeding of a youngster. In the first picture the young penguin approaches the adult, in the second the adult leans down with open bill, and finally the youngster inserts its bill into that of the adult and is fed with food held in the crop.

expedition in the *Belgica* wrote:

> Imagine a little man, standing erect, provided with two broad paddles instead of arms, with head small in comparison with the plump stout body; imagine this creature with his back covered with a black coat . . . tapering behind to a pointed tail that drags on the ground, and adorned in front with a glossy white breast-plate. Have this creature walk on his two feet, and give him at the same time a droll little waddle, and a pert movement of the head; you have before you something irresistibly attractive and comical.

During the late nineteenth and early twentieth century the unfortunate penguins had far more to worry about than the possibility of being eaten by marooned explorers. They have a thick layer of fat beneath the skin (two centimetres thick in the case of the King Penguin) and this was eagerly exploited by the simple expedient of boiling the penguins down in iron try-pots. The King Penguins on South Georgia, for example, were boiled for oil in large numbers by traders. The most infamous penguin oil factories of all were on the Macquarie Island group. In 1891 a certain Jospeh Hatch was granted a lease by the New Zealand government to collect oil. His highly successful operation lasted for more than 25 years before he finally had to give up because of public opposition. Hatch concentrated at first on the King Penguins, but later turned his attentions to the smaller but far more abundant Royal Penguins. His 'season' was a period of six weeks in February and March, during which something like 150,000 birds were processed. As it happened, these operations had little effect on the population since the numbers taken did not exceed the annual increase.

This close-up of an adult Chinstrap Penguin shows well the colour and pattern of the plumage. The narrow band of black-tipped feathers under the chin give this species a distinctive appearance. Chinstraps are the characteristic penguins of the Maritime Antarctic.

There are many gaps in our knowledge of penguin life-histories, and precious little is known about their underwater activities. By far the greater proportion of their time is spent at sea, but most of the information on them refers to the period spent on land.

Three species in the genus *Pygoscelis* breed in the Antarctic, and of these the Adelie Penguin has a circumpolar distribution on the Antarctic Continent and is the most southerly breeding of the three. The Gentoo Penguin is the most northerly species, actually breeding as far north as the Falkland Islands, but reaching 65°S on the Antarctic Peninsula, where the southernmost colony appears to be that on Petermann Island (65°10′S, 64°10′W). The Chinstrap Penguin has a restricted longitudinal distribution, breeding on the Antarctic Peninsula and islands in the southern Atlantic Ocean, although there is evidence that it may be extending its range. All three species overlap in the Antarctic Peninsula area (55°–65°S), but where they share the same rookeries they normally segregate by species in pure colonies. The two largest living species of penguin are the King Penguin and the Emperor Penguin. The former has a circumpolar distribution on islands between latitudes 52°S and 55°S, and all known breeding colonies lie north of the normal maximum limit of the pack-ice. On the other hand the Emperor Penguin, perhaps the most remarkable bird in the world, breeds exclusively along the edge of the Antarctic Continent and its ice shelves.

Three other species – Rockhopper, Macaroni and Royal – are crested penguins of the genus *Eudyptes*. The species in this genus breed on warm temperate islands north of the Antarctic Convergence, but in the case of the above species extend south to the subantarctic islands, but only the Macaroni Penguin has penetrated as far south as the cold islands of the Maritime Antarctic. There is some evidence that the Macaroni Penguin is trying to extend its range further south. In the summer of 1979-80 Polish scientists found a few nests on King George Island in the South Shetlands, and there may also be a colony at Macaroni Point on Deception Island, also in the South Shetlands. Early in March 1979 a female of this species was captured on Humble Island near Palmer Station (64°46′S, 64°03′W) on the Antarctic Peninsula. The Rockhopper Penguin on the other hand breeds on half a dozen subantarctic islands and gets no further south than South Georgia where it was first recorded breeding in 1975, but it is very rare there. Even more restricted is the Royal Penguin which, as already noted, is endemic to Macquarie Island.

The area supports enormous numbers of penguins, although some of the earlier population figures were based on inspired

An Adelie Penguin incubating its eggs. Note the rudimentary nest constructed of flattened stones, just adequate enough to form a shallow cup.

Right *A Chinstrap Penguin with its egg on a boulder-strewn beach. The normal clutch is two, but sometimes only one egg is laid.*

Overleaf *A pair of Chinstrap Penguins with two young, one of which is being fed. Once the young moult they are almost indistinguishable from the adults.*

guesswork rather than proper census work. Nevertheless, calculations, admittedly rather crude, suggest that penguins as a whole constitute around 65 per cent of the total seabird population of the Antarctic. There is now evidence to show that penguin numbers have increased markedly at many breeding colonies over the past few decades. Several factors have played a part in this, and relief from the nineteenth and early twentieth century persecution for oil is a major factor so far as the King Penguin is concerned. Additional factors which have also benefited other species, are an increased food supply following depletion of the whale populations, and climatic changes which have not only resulted in more ice-free areas, but have probably also reduced the mortality rate.

The Adelie Penguin population of the Antarctic Peninsula and the islands of the Scotia Arc has recently been reviewed by biologists of the British Antarctic Survey. The general picture that has emerged is of a gradual, but continuing increase at the more northerly colonies, and a generally slower increase (if any at all) at the more southerly sites. Some of the colonies on the Antarctic Continent have actually decreased in numbers possibly due to disturbance from nearby bases. What the available data indicate is that by far the greatest increase in numbers has occurred in the Adelie and Chinstrap Penguins which feed predominantly on krill. Although the Gentoo Penguin also feeds extensively on krill, it takes significantly more fish than the other two species, and has increased less markedly. The most notable increases have been in the populations of the Scotia Arc islands and South Georgia, areas where the whalers were most active in the early decades of the present century.

The large penguin rookeries must rank among the great natural sights of the Antarctic. The numbers of Adelie Penguins

Gentoo Penguin with young.

are quite enormous, although not always in large rookeries. Cape Crozier is the site of a rookery which was already in existence there when the first explorers arrived on the scene. However, there does appear to have been a decrease in the numbers of Adelies at this location. In the 1964-5 season it was estimated that there were 175,000 breeding pairs (producing around 157,000 chicks), but by 1970-1 there were only about 105,000 breeding pairs. Similar decreases have occurred in the colonies at Cape Hallett and Cape Royds. It is perhaps significant that these colonies are more or less on the southern fringe of the range of the Adelie Penguin. From Cape Crozier to Cape Adare in the western Ross Sea there are about 1100 kilometres of coastline with about two dozen colonies which at one time were estimated to contain 600,000 breeding pairs. Although the colonies are almost invariably in coastal areas, there was an interesting exception a few years ago when a new colony was discovered on the Antarctic Continent at Schirmacher Ponds (70°45′S, 90°35′E), a freshwater lake system some 100 kilometres from the nearest open sea.

Some of the biggest populations of Adelie Penguins are found on the islands of the Maritime Antarctic, and the numbers have been increasing. In the South Orkney Islands, for example, it has been estimated that some five million nest on Laurie Island alone. On Signy Island the breeding population rose from about 10,500 pairs in 1947-8, to 31,525 pairs in 1978-9. One of the largest continuous-breeding colonies known in the early 1970s was at Stranger Point on King George Island in the South Shetlands. This had an estimated 8000–10,000 nests, and birds breeding in the centre of the colony had to cross some 90 territories (nests) in order to reach their own nests.

Overleaf *Gentoo Penguins at Paradise Bay. Note how the two nests in the foreground are spaced just beyond pecking distance.*

Right *Gentoo Penguin exhibiting aggressive behaviour to a young bird, probably the progeny of a neighbouring pair.*

Obviously it is almost impossible to know how long some of the Adelie Penguin colonies had been in existence before the arrival of the first explorers. However, in January 1959 five small colonies were discovered on the Balleny Islands. Examination and carbon dating of the 65–95-centimetre deposit of guano at one of these colonies (on Beufort Island) gave an estimated age of 2000 years.

The Chinstrap Penguin nests in large colonies which may contain tens or hundreds of thousands of birds. In the South Shetland Islands, Zavodovski Island alone may have around 14 million nesting birds. In 1970-1 the breeding population on the Elephant Island group, also in the South Shetlands, was estimated at 151,600 pairs. In the South Orkneys there may be more than one million Chinstraps on Laurie and Saddle Islands. The present breeding population on Signy Island is around 87,500 pairs. The population on the Gourlay Peninsula on Signy Island grew from a little under 10,000 pairs in 1950-1 to more than 14,000 pairs in 1969, a 40 per cent increase in 18 years. Another census there in November 1978 produced a count of 13,063 pairs. At South Georgia, the northern limit of breeding for this species, the population is currently estimated to be between 2000 and 6200 pairs.

So far as the Gentoo Penguin is concerned there is not a great deal of data to go on. The latest estimate of the South Georgia breeding population is 100,000 pairs, and for the Marion and Prince Edward Islands 1500 pairs. There appear to be no comparable figures for the other breeding locations of this species.

When we come to look at the three species of crested penguin, we find that there is almost no data for the Rockhopper Penguin. On Marion Island in the late 1960s one observer

estimated breeding populations of 250,000 pairs at Bullard Beach and about 500,000 pairs at Kildalkey Bay. However, an estimate for the mid-1970s gives the *total* breeding population of this species on the Marion and Prince Edward Islands as only 128,300 pairs. The apparent discrepancy merely indicates the difficulties of counting penguins in the breeding colonies, rather than a real population change. An earlier estimate for the Îles Crozet gave 157,000 breeding pairs. In the case of the Macaroni Penguin, the largest proportion of the world population is on South Georgia where, fortunately, there are very recent data suggesting a breeding population of 5,440,000 pairs. The species is known to have increased considerably on the Marion and Prince Edward Islands in the 15 years up to 1970, and in the mid-1970s the total breeding population was estimated at 467,000 pairs. Other important population estimates for the Macaroni Penguin are 455,000 pairs on Îles Crozet and one million pairs on Îles Kerguelen. The Royal Penguin population on Macquarie Island may total two million birds, of which over 500,000 are in a colony which covers an area of 6.5 hectares at Hurd Point.

Finally we have the King and Emperor Penguins whose populations are small compared with the species discussed above. There is pretty clear evidence that the King Penguin is increasing in most, very possibly all, its breeding localities. A survey on South Georgia in the 1970s discovered numerous colonies which had not been reported previously. It would seem that the island has an expanding population with new small colonies being established by overspill from the major breeding colonies. The *total* population (i.e. including all age classes) on South Georgia in 1946 was estimated at over 12,000, and the present total at just over 57,000 birds. It is not yet clear whether the King Penguin on South Georgia breeds twice every three years, or biennially as is said to be the case on Îles Crozet. If the former is the case then the total *breeding* population at South Georgia is probably about 22,000 pairs, of which 15,000 pairs may breed annually. The major proportion of the world population of this species is in fact on Îles Crozet for which a figure of 455,000 pairs has been quoted, but whether this represents the annual or total breeding population is not clear.

Anything like an accurate population estimate for the Emperor Penguin is difficult to come by, which is hardly surprising in view of the very remarkable habits of this species, and the fact that the populations are known to fluctuate strongly. At Cape Crozier for example, high adult mortality and extreme weather conditions have reduced the numbers. Over 30 colonies have so far been discovered, some of them only recently. They vary in size from 150 pairs (on the Dion Islands off the Antarctic Peninsula) to some 50,000 pairs on Coulman Island. The total population of this strictly antarctic species is probably something over 300,000 birds.

Having dealt with the question of the actual numbers of penguins, we can now look at some aspects of their life history during the breeding period. The Adelie Penguin is probably the most sociable, and some would say the most comical of the penguins. While there have been instances of Adelies nesting at considerable heights above sea level, they tend to nest mainly on raised beaches in areas where the sea-ice opens early in spring and can normally be relied upon to disperse locally in summer. When they first return to the breeding grounds in October the spring thaw has usually not yet begun. At some locations, such as Cape Adare, Cape Hallett and Cape Royds, the returning Adelies may face a journey of up to two weeks

A breeding colony of Rockhopper Penguins on sloping, boulder-strewn ground – a very typical site. Although two eggs are normally laid, the second rarely hatches.

Rockhopper Penguins returning to the nesting colony from the sea. Whereas most penguins walk or stroll, this species – like the Macaroni – hops.

Above *A pair of Rockhopper Penguins change incubation duties at the nest; this pair have only one egg.*

Below *This Rockhopper Penguin is in the characteristic hopping posture with the flippers back and the head forward.*

duration over the sea-ice, across which they march in seemingly endless files to the site of the rookery. Incubating through the cold conditions of early spring, they rear their chicks in the brief summer and launch them upon the world while the summer flush of plankton remains. The Adelie Penguin is the best adapted of the three *Pygoscelis* species for life in conditions of extreme cold, and it needs to be.

Nearly 50 years ago the famous American ornithologist Robert Cushman Murphy, reported how 'the Adelies frequently get snowed under during spring and summer blizzards, and they have been known to live for weeks beneath the crust before they thawed out. No harm necessarily comes to the eggs from this.' As a species the Adelie Penguin is confronted by two dreadful possibilities: that the pack-ice will not disperse sufficiently for foraging by the adults, and that the sea will freeze again before the young are ready to leave the colonies. There is indeed evidence that such events occur occasionally, for the hollows and recesses of various coastal and inland rookeries have been reported as being 'dense with the bodies of almost mature chicks, dehydrated but not decomposed.' Even under reasonable conditions the parent birds may have to travel great distances in search of food, and studies at Cape Crozier indicated that each pair of adults made 35–40 feeding trips during the fledging period. In years characterized by very bad sea-ice, the breeding success of the Adelie Penguin may be very low or even non-existent. At the high southern latitudes where this species breeds, ice-free ground is at a premium, so the birds utilize such areas very efficiently and their nests of pebbles are in close proximity.

Where two or more species of pygoscelid penguin are found together, a typical partitioning of the ice-free areas usually

prevails, reflecting their different habitat preferences and patterns of colonization. The Adelie Penguins select knolls or ridges, the Gentoo Penguins nest on low flat areas where their large nests of pebbles are widely spaced, the Chinstrap Penguin prefers rough boulder-strewn slopes at the higher elevations, and on occasion even practises true cliff nesting.

Obviously there are other methods by which these penguins achieve ecological separation. At Signy Island in the South Orkneys, for example, research has shown that while the abundant Adelie and Chinstrap Penguins both feed on the krill *Euphausia superba*, there is a significant difference in the sizes of their prey. Although the Chinstrap's general pattern of breeding is similar to that of the Adelie Penguin, there are differences in timing and in the length of incubation spans. The extended incubation span of the early-laying Adelie Penguin allows it to cross fast-ice or pack-ice in its search for food, and to travel further from land and spend more time at sea than the Chinstrap Penguin. On Signy Island the latter begins breeding some three weeks later than the Adelie Penguin. On Deception Island in the South Shetlands, where volcanic activity may warm the ground locally, Chinstrap Penguins have been found to lay earlier than elsewhere. On Signy Island the first incubation period by the male Adelie Penguins was from 7–18 days (average period 11 days), while in the case of the Chinstrap Penguin it ranged from 1–9 days with an average of about three days. Since the male penguins arrive first at the rookeries and normally take the first spell of incubation, they have to fast and live on their fat reserves until relieved by the female. In the case of the Adelie Penguins on Signy Island, the period of fasting proved to be about 40 days, during which they lost about one-third in weight.

Above *A returning Rockhopper Penguin begins the greeting ritual. The Chinstrap, Gentoo and Rockhopper Penguins all have an elaborate nest-relief ceremony.*

Below *This Rockhopper Penguin is incubating a clutch of two eggs.*

A pair of Rockhopper Penguins sit peacefully in the rookery.

Whereas Gentoo Penguins in the higher latitudes nest on bare hillsides and beaches, those on the subantarctic islands do so amongst tussock grass, sometimes well away from the sea. On the Marion and Prince Edward Islands the breeding colonies are invariably on a hill or ridge, avoiding the beaches with their concentrations of King and Macaroni Penguins. Although the Gentoo Penguin lays two eggs, at both Marion Island and Îles Crozet it was noted that hardly any pairs managed to fledge more than one chick. The nests of this penguin tend to be widely spaced, and studies at four different colonies showed an average distance between nests of just over 103 centimetres.

The Chinstrap, Gentoo and Rockhopper Penguins all have an elaborate nest relief ceremony which consists of a number of separate displays, each of which is performed one or several times. These displays are most similar in the Adelie and Chinstrap Penguins. There is the 'loud mutual display' in which both birds stand and wave their necks back and forth, at the same time cackling with open bill. In the similar 'quiet mutual display' both birds utter a soft humming sound with the bill closed. Then there is a display known as 'circling' in which one birds walks around the rim of its nest while nodding its head. The Gentoo Penguin diverges from this pattern in that its loud mutual display lacks the neck-waving, and the sound uttered can be likened to the braying of a donkey. The Gentoo Penguin also has a second main display called the 'bow-gape-hiss' in which it bends down to the nest, opens the bill to show the bright red lining, and hisses.

When we come to consider the breeding of the Macaroni and Royal Penguins we find that the really large colonies are invariably sited on level or gently sloping ground. The Rockhopper Penguin has a similar preference, but is not averse to nesting on steep slopes consolidated by tussock grass or other vegetation, where they build up the soil on the downhill side to form a nesting platform. They usually nest in this way when their colonies adjoin those of the larger Macaroni Penguin. Where two species of this group occur on the same island, the larger species arrives before the smaller and is the first to lay. On Macquarie Island the mean laying date of the Royal Penguin is nearly a month before that of the Rockhoppers, so their chicks leave for the sea that much earlier. The staggering of the breeding season in this way means that two great armies of young birds are not competing for food simultaneously.

On Macquarie Island the great rookeries of the Royal Penguin are some way from the sea, and on landing on the beaches some individuals are faced with a journey of as much as three kilometres or more, and have to climb to an elevation of some 150 metres above sea level, negotiating boulders in the process. A nice description of the sort of journey involved is given by Dr Mary Gillham in her book *Sub-Antarctic Sanctuary: Summertime on Macquarie Island* (1967):

> There was a comical air of urgency about the birds as they scrambled up the sides of the road blocks, often slipping back and having to try again. A brief pause on the summit and they skated down the further side on their tails – a move which never went quite according to plan, to judge by the inelegant attitudes and astonished expressions. If a bird lingered to rest or preen there was the inevitable traffic jam until the offender was jostled out of the way by the press of bodies behind.

It is a few hundred metres from the coast that the bordering tussock grasses open out into a huge rookery, where it is not

only unbelievably noisy (as are most penguin rookeries when both sexes are present), but also distinctly unsalubrious. So densely packed are the birds that the vegetation has largely disappeared and the adults incubate and rear their chicks in a deplorable state of filth. Extensive soil erosion is a feature of many large penguin rookeries. The total breeding span of the Royal Penguin on Macquarie Island is some 243 days, compared to a span of only 193 days for the very closely related Macaroni Penguins on Heard Island. The shorter cycle on Heard Island is no doubt related to better feeding conditions and longer days.

The three species of crested penguin are notable in normally laying two eggs of which the first is often smaller than the second and is usually lost early in incubation so that only a single chick is reared. Recent studies on Macaroni and Rockhopper Penguins on Marion Island showed that both eggs hatched at only six per cent of Rockhopper nests and at no Macaroni nests. Even when both eggs did hatch one chick died of starvation within 12 days.

Last, we come to the two most remarkable of all the penguins, the King and the Emperor. They are similar in appearance, but the King Penguin is shorter and only about half the weight of the Emperor Penguin, and has brighter yellow coloration. Neither species constructs a nest, the single large egg during incubation being carried on top of the feet and warmed by a fold of feathered skin of the belly. Here the similarities largely cease, and in other aspects of their life history the two species differ considerably.

King Penguin rookeries are almost invariably on low, bare ground, often in the shelter of dense tussock grass. At Paul Beach, South Georgia, winding paths have been trodden through

Above left *Macaroni Penguins at a breeding colony on South Georgia, with young birds visible on the right of the picture.*

A Macaroni Penguin surrounded by incubating Rockhoppers in the rookery.

Overleaf *Royal and King Penguins on Macquarie Island. The Royals are making their way up a steep slope to their rookery.*

Above *Royal and King Penguins on the beach at Macquarie Island appear completely unconcerned at the close proximity of the tourists.*

Right *A Royal Penguin in moult showing the large bill and white (rarely black) cheeks and throat.*

Overleaf *Royal Penguins amongst tussock grass on Macquarie Island. The population of this species on the island may total two million birds.*

the tussock grass by the penguins. On Heard Island, the northern Spit Bay colony is sited several hundred metres from the sea, on well-drained and compacted earth in a broad valley of tussock grass and Azorella. Elephant seals and fur seals may often be found in close proximity to King Penguins, but seem to bother the penguins not at all. All the rookeries known at present lie north of the northern limit of the pack-ice, as mentioned earlier. This relationship between ice distribution and breeding sites is very clear, and the reason for it is that because young King Penguins remain in the colony throughout the winter months and do not rely entirely on their fat reserves, the adults must have access to the rookery all through the year, and the presence of ice would obviously impede such access.

In contrast to the highly synchronized breeding activities of many antarctic birds, the King Penguin has an exceptionally long (14–18 months) and rather strange breeding cycle. Before breeding occurs the penguins, fat from feeding in the rich offshore waters, assemble on shore to moult, a process which occupies some 4–5 weeks during which they take no food. Once the moult is complete they go back to sea and spend a further 2–3 weeks fattening up again. Courtship begins in October, the first eggs are laid early in November, and the first chicks hatch out in December. The chick is fed from the time it is hatched until it is 10–13 months old. Because of this spread of the breeding cycle any individual bird must breed two months later in each successive season, thus normally completing two cycles every three years. What normally happens is that there is an 'early' year when laying takes place from November to January, then a 'late' year when they lay from February to April, followed by a year when they are 'ineffective' breeders.

Above *Porpoising Royal Penguins approach a wide expanse of beach on Macquarie Island.*

Left *A solitary Royal Penguin stands surrounded by Gentoo Penguins on Macquarie Island.*

Given the sequence of events outlined above, it is obvious that the chicks cannot reach full size before winter arrives, and so they remain in the colony throughout the gruelling winter months. During this time they commonly huddle together in enormous crèches, and are fed by the parents at irregular intervals, sometimes only every third or fourth week. Not surprisingly, all the chicks lose body weight (something of the order of 40 per cent) and many of the smaller ones die of starvation during the coldest months of June, July and August. However, the survivors gain weight rapidly when food becomes more plentiful in September, and by October or November they are moulting and leaving for the sea. Because of this complex breeding cycle, a January or February visit to a breeding colony reveals all stages of breeding activities – courtship, incubation, brooding, chick-feeding, and moulting – and newly

A pair of King Penguins resting on the beach.

hatched chicks are in company with those hatched a year earlier.

Strange though the way of life of the King Penguin is, that of the Emperor Penguin verges on the unbelievable. Among birds it is a record breaker: it is the only bird that breeds exclusively along the edge of the Antarctic Continent, it is the only bird that breeds elsewhere than on land, and it is the only bird that lays its egg in the antarctic winter. There is in fact no other species of higher animal that breeds in such forbidding circumstances. Known to the earlier voyagers only as a summer wanderer among the pack-ice, its breeding haunts and habits remained a mystery until October 1902 when the first rookery was discovered at Cape Crozier. This rookery was visited then by Dr Edward Wilson of the British Antarctic Expedition of 1901-4. By a careful count of deserted eggs and chicks he found that the mortality of the year's brood was no less than a

Left *This close-up of a King Penguin shows the characteristic comma-shaped ear patch and the beautiful texture of the plumage.*

Above *A pair of King Penguins in a thoughtful mood.*

staggering 77 per cent. He watched the extraordinary migration, on drifting ice-floes, of the adults and down-covered chicks, and concluded that the habits of this bird were 'eccentric to a degree rarely met with even in ornithology'.

Wilson's return visit to this rookery in 1910 together with Apsley Cherry-Garrard and Lt H.R. Bowers is graphically described in Cherry-Garrard's book *The Worst Journey in the World* (1922). The purpose of their journey was to watch the winter hatching of the Emperor Penguins, and it is doubtful whether any men since have subjected themselves to such incredible suffering in the pursuit of scientific knowledge.

The Emperor Penguin lays its egg on the sea-ice in May or June just as winter is beginning, and when they hatch some two months later the chick is greeted by a world of perpetual darkness in which the temperature may drop below −40°C, hurricane force winds blow persistently, and there is no food. The eggs are incubated for 60–64 days, only by the males who huddle together in tight formations known as *tortues* or huddles, which may contain hundreds and even thousands of birds. Huddles form on the flat surface of the sea-ice and the incubating birds are able to shuffle from place to place with the egg in position on top of the feet. Huddling, although a protection against the extreme weather conditions, can lead to losses of eggs and chicks through squabbling and inevitable accidents. When the cold winds blow the birds pack together as tightly as possible in the customary standing position, all facing towards the centre and leaning slightly inwards with beaks resting on the shoulders of the birds in front. Individuals on the windward side of the huddle lose no opportunity of trying to force their way into the middle, or failing this of leaving the windward side altogether to walk round and rejoin in the shelter provided by the rest of

An Emperor Penguin stands with its chicks on the ice.

This picture of a large gathering of Emperor Penguins on the ice, although taken on a sunny day, illustrates well the exposed environment where these remarkable birds breed.

the pack.

Despite the squabbling that may occur in the huddles, the Emperor Penguin exhibits a degree of communal behaviour that is unique among penguins. When the chicks have hatched and are being brooded by the adults (for another 40 days) the air may resound with their piping calls which one writer has described as the sweetest natural sound to be heard in Antarctica. The very size of the Emperor chick means that it simply cannot reach maturity in the brief summer months, it would be an impossibility for the adults to supply enough food. So it is the end of the year before they finally moult into juvenile plumage and depart for the open sea. The females return to relieve the emaciated males about the time the chicks hatch, and the males then leave to feed at sea for several weeks and regain weight.

It is possible that the Emperor Penguin holds yet another record – that of diving deeper than any other bird, for underwater observations and experiments at holes in the ice off Cape Crozier have shown that they can stay below for at least 18 minutes, and reach a depth of 265 metres.

Although penguins are essentially birds of the sea and coasts they do occasionally turn up in unexpected places. For example, on 31 December 1957, the tracks of what was apparently an Emperor Penguin were seen by a party traversing the Antarctic Continent over 400 kilometres from the nearest known sea, and on 1 January 1958 a different party hundreds of miles distant came across tracks, almost certainly those of an Adelie Penguin, at an altitude of just over 1400 metres and more than 300 kilometres from the nearest known sea. These tracks were heading in the direction of the South Pole!

Albatrosses

For those approaching the Antarctic by ship, perhaps the most beautiful and exciting seabirds of the southern seas are the albatrosses, whose grace and powers of flight have been commented upon by voyagers ever since ships first sailed south into these cold waters. They were generally referred to by sailors as gooneys or mollymawks, but the latter term is normally now used only when referring collectively to the smaller species such as the Black-browed Albatross *Diomedea melanophris*, Shy or White-capped Albatross *D.cauta*, Yellow-nosed Albatross *D.chlororhynchos* and Grey-headed Albatross *D.chrysostoma.* There are 13 species of albatross in the world of which one breeds in the tropics (on the Galapagos Islands) and three in the north Pacific. The other nine species inhabit the world's windiest latitudes between the Tropic of Capricorn and the Antarctic Circle, and most of them south of latitude 40°S. Within the area with which we are concerned six species of albatross breed, and those only on the subantarctic islands (see Appendix 2). Two other species which breed only in Australian and New Zealand waters – the Shy Albatross and the Royal Albatross *D.epomophora* – occasionally wander south into the Subantarctic. The Royal Albatross may, in fact, be a more frequent visitor than the records suggest, since adults at sea can easily be confused with those of the Wandering Albatross *D.exulans.* These two species are the largest living albatrosses and have a wing span that may approach four metres.

There is a certain amount of superstition attached to the albatross, or at least to some species; who has not heard of Coleridge's *The Rime of the Ancient Mariner* which appeared in 1798 and immortalized the albatross?

> And I had done a hellish thing,
> And it would work 'em woe:
> For all averr'd I had killed the bird
> That made the breeze to blow.
> Ah wretch! said they, the bird to slay,
> That made the breeze to blow!

The illustrations to Coleridge's literary gem apparently portray the Wandering Albatross, but there is in fact no bad omen to be attached to the killing of this great albatross since the species involved in the superstition is actually the Sooty Albatross *Phoebetria fusca.* The old sailors thought that this albatross was the reincarnation of sailors who had fallen overboard in storms and been drowned. A good example of this superstitious belief is given by L. Harrison Matthews in his book *Wandering Albatross* (1951):

Right *Royal Albatross in close-up showing the tube nose.*

Below *Royal Albatross and young in tussock grass.*

A Shy or White-capped Albatross lands on the sea and prepares to fold its long wings.

Hermann watched him gliding alongside the weather leach of the main lower topsail just above the main yard-arm, and looking down over the bulging belly of the course. 'See what he's doing?' he said. 'He's having a look at the bunt-line and clew-line lizards to see they're all clear for running and won't foul if we have to clew up in a hurry for the next squall; they always do that. I wonder how long ago he went overboard.'

In actual fact the belief of bad luck attending the killing of an albatross, which gained such wide acceptance, seems to have absolutely no origin at all in the many fables of the sea. The famous *Rime* was entirely the product of Coleridge's vivid imagination. In any case it has recently been suggested that the bird concerned was not an albatross at all, but a giant petrel.

Superstition and general interest notwithstanding, the early mariners, along with whalers and sealers, never had the slightest compunction in killing albatrosses and using them for quite mundane purposes. Sailors sometimes made feather rugs from albatross skins, this no doubt being the origin of the name 'Cape Sheep' that was sometimes applied to them. Whalers and sealers in the Antarctic often ate albatrosses and, on the principle of 'waste not want not', used their bones for making needles and other implements, their feathers for clothing, and the webbed feet for tobacco pouches. Sailors also used albatrosses to supplement their normal, somewhat monotonous diet when opportunity offered. Take, for example, an entry in Captain Cook's journal for November 1772:

Tuesday 24th. Winds NW, SW to SE. Course S12°E. Distce sail'd 50 miles. Latd in South 35°25′, Longd in East of Greenwich 17°44′. Longd made from the Cape of Good

A Shy Albatross building up speed preparing to take off from the sea.

> Hope 0° 39′W . . . Many Albatrosses about the ship, some of which we caught with Hook and line and were not thought dispicable food even at a time when all hands were served fresh Mutton.

There was a well-developed technique of 'fishing' for albatrosses which the early sailors used with great success. It has been described by various writers, but let us refer again to Matthews (1951):

> This is the way we took them in a sailing ship, he added, cutting a triangular piece, about four inches along each side, from the copper. Then he cut another piece out of the centre, so that a triangular frame was left, its sides about half an inch wide, and in one corner he tied a lump of fat salt-pork from the harness cask. To the base opposite the pork he tied a cod-line, and prepared for his fishing by throwing out a handful of pork chunks as ground bait to attract the birds. They gathered round, and then he slipped his triangle over the side and paid out the line until it drifted in among them; several picked it up but dropped it as soon as they felt the hard copper, before Hermann had the line adjusted right and ready to snare one properly.
>
> Then as one approached the floating lure he drew it gently towards him, and away from the bird. It stretched out its neck as it paddled after, and the moment it grabbed the pork Hermann jerked the line and it was caught fast, the hooked tip of its beak jamming between the limbs at the point of the hollow triangle, whose ragged edges bit into the horny bill and held it firm.

A Wandering Albatross in characteristic effortless flight over the Southern Ocean.

The art of gliding and soaring is exhibited to the very peak of perfection by the albatrosses who spend the greater part of their lives airborne over the oceans. Their flight is so perfected that they expend the least amount of energy in travelling vast distances in search of their fish, squid and plankton food. Exactly how far adult albatrosses range for food during the breeding season is not known in detail (see below), but certainly several hundred kilometres would be no obstacle. It is known, however, that between breeding seasons some Wandering Albatrosses return precisely to known feeding areas. It is the turbulent uplift of air deflected from the windward surface of the waves that provides the motive power for the flight of the albatross. They require a minimum windspeed of about 16 kilometres per hour in order to sustain gliding flight, and they are able to take advantage of the rising warm air from a ship in motion to maintain flight; often they 'hang' over the ship's stern. The wind does most of the work for the albatross. Watch them in flight and it will be seen that they gather speed by gliding down-wind towards the surface of the waves, then shoot upwards into the wind losing ground speed as they quickly gain height before falling away in another long down-wind glide.

Albatrosses rarely penetrate beyond the outer edge of the pack-ice, a point noted by many early observers. The reason why the Wandering Albatross and the mollymawks do not normally cross the frontier between the sea and the ice is quite obvious. It is simply that the conditions on which they depend for easy flight hardly exist beyond it. The great slabs of ice suppress the waves and flatten the sea except for a slight swell; there is no turbulent uplift. Also, the westerly wind belt stops at the sea/ice frontier, which in fact it creates by arresting the northward drift of the sea-ice.

Here the courtship display of a pair of Wandering Albatrosses reaches its climax when, with bills pointed upwards, wings outstretched and chests almost touching, both birds shriek and yell.

Nevertheless, there is one species of albatross that not only passes this frontier to be found within the ring of pack-ice, but also ranges along the shores of the Antarctic Continent itself. This is the Light-mantled Sooty Albatross *Phoebetria palpebrata* which is completely circumpolar and has the most southerly range of any albatross. The two species of Sooty Albatross are specialized for buoyant flight as few other species are, being surpassed in this respect only by the frigate birds of the tropics.

Most of the albatrosses are great travellers, and none more so than the Wandering Albatross which in the course of its life may circumnavigate the world from west to east south of latitude 30°S again and again carried by the west winds. This happens because they return to their nesting sites by continuing in the direction of their departure, thus making the circuit of the earth. Wandering Albatrosses ringed on their breeding grounds in South Georgia, Îles Kerguelen and Îles Crozet, have been recovered along the south-east coast of Australia, and to a lesser extent off the coasts of South Africa and South America. Wandering Albatrosses (as well as other species) frequently followed ships for the garbage thrown overboard and the organic material brought to the surface by the disturbance caused by their passage. However, the high speed of modern ships means that this behaviour is now less in evidence. Black-browed Albatrosses have been recorded way up in the northern hemisphere on a number of occasions, reaching as far north as Spitzbergen. One spent 34 years up to 1874 at a gannetry in the Faeroes, and another appeared at a gannetry at Hermaness in the Shetland Isles in 1974 and was still to be seen there in 1981.

The close study of albatrosses presents many obvious problems. They breed mostly on remote islands where, in the Antarctic, bad weather and difficulties of access plague the

A Light-mantled Sooty Albatross in flight. This species has the most southerly distribution of any albatross.

would-be human observer. They come to land only to breed, and a great part of their life is spent over the open oceans. So far as this aspect of their life history is concerned there are still considerable gaps in our knowledge. The study of albatrosses on the breeding grounds is essentially a long-term affair, not the least of the problems being that the breeding cycle of certain species is so long that they can nest only biennially when successful. Add to this a life expectancy measured in decades, and the fact that the larger species do not breed at all until four to six years of age at least, and often not until 9–10 years, and the difficulties are very apparent. Nevertheless, thanks to the sterling work of biologists of several nationalities in the Antarctic, there is now a considerable volume of knowledge.

How many breeding pairs of albatrosses are there in the Antarctic? That in itself is a question that is not as easy to answer as may be supposed, for even taking a census of breeding albatrosses has its problems. In the case of the Wandering and Grey-headed Albatrosses, for example, which breed only biennially when successful, there will in any one season always be a substantial proportion of the *total* breeding population away at sea. However, taking *total* breeding populations of all six antarctic species we can arrive at a very approximate minimum total of 163,600–178,300 pairs. In order of abundance the total breaks down as follows: Grey-headed Albatross 67,600 pairs, Black-browed Albatross 61,500–70,500 pairs, Wandering Albatross 16,900 pairs, Light-mantled Sooty Albatross 8900–14,600 pairs, Yellow-nosed Albatross 5000 pairs, and Sooty Albatross 3700 pairs. However, in the case of the Light-mantled Sooty Albatross the figure quoted is certainly far too low since there are apparently 'thousands' breeding at

Îles Kerguelen. In the case of the two biennially breeding species – the Wandering and Grey-headed – the approximate *annual* breeding populations are over 8800 and 42,500 pairs respectively.

With the exception of the Sooty and Light-mantled Sooty Albatrosses which are solitary nesters (or at most in small groups) on steep cliffs and precipices, the breeding albatrosses of the Antarctic are colonial in their nesting habits, and some of the colonies contain thousands of closely packed birds. The nests of the Wandering Albatross, however, are widely scattered. For example, at one breeding site of this species it was found that there were 106 nests in an area of one hectare. There is no doubt that watching albatrosses at sea demonstrating their mastery of the elements is a fascinating experience. Equally, if not more fascinating, is a visit to a breeding colony on a remote island, where one can sit so close to these magnificent and fearless birds that every subtle detail of the beauty of their plumage can be savoured to the full.

All the albatross species breeding in the Antarctic build a definite nest of compacted earth and vegetation, with a hollow cup on top. They defend the immediate vicinity of this nest, but are otherwise not territorial. To the uninitiated observer a breeding colony of thousands of albatrosses may appear somewhat chaotic with the nests sited just any old how; but this is not so and a high degree of order prevails. Studies made at a colony which contained large numbers of both Black-browed and Grey-headed Albatrosses revealed that the nests of the former were pretty consistently spaced at 1.55 metres apart, whilst those of the Grey-headed (even when nesting cheek by jowl with the Black-browed) were 1.31 metres apart. There is ample evidence that colonial seabirds are influenced by the level of social activity within the colony and that this serves an important function, so much so that breeding success may be very low in small colonies. This was well demonstrated on Macquarie Island where the two species just mentioned have quite small populations (only some 38 pairs of Grey-headed Albatrosses in the 1978-9 season), when at one colony of only 12 pairs only one young was reared in five years.

South Georgia is one locality where both the Black-browed and Grey-headed Albatrosses breed in considerable numbers, and where their ecology has been studied in some detail by biologists of the British Antarctic Survey. These studies have shown that although the timing and duration of the breeding cycle of both species are similar, the Grey-headed Albatross lays earlier after a longer pre-laying attendance period, and its chick takes 25 days longer to fledge. Because of the longer breeding cycle the Grey-headed Albatross can breed only biennially when it succeeds in rearing young. On the other hand its adult life expectancy is greater than that of the annually breeding Black-browed Albatross.

There are also food differences between the two species. Krill predominates (by weight) in the diet of the Black-browed Albatross which is thus able to rear its young more quickly than can the Grey-headed Albatross, whose young are fed predominantly on less nutritious squid. The estimated foraging range of both species from South Georgia is just over 900 kilometres, compared to an estimated foraging range for the Wandering Albatross of 2650 kilometres. The extent to which a failure in the food supply during the nesting period can have serious consequences was very apparent at South Georgia in the 1977-8 breeding season. Commercial fishing operators were unable to locate krill swarms around South Georgia that season and there was an unprecedented failure by Black-browed

Above *A beautiful Grey-headed Albatross on an elevated nest which has probably been in use for several seasons.*

Left *In the magnificent setting of South Georgia a Light-mantled Sooty Albatross broods its youngster.*

Overleaf *A Black-browed Albatross landing near its nest.*

A typical nesting colony of Black-browed Albatrosses amongst tussock grass.

Albatrosses and Gentoo Penguins to raise their chicks. On the other hand the mainly squid-feeding Grey-headed Albatrosses had a very good breeding season.

All the albatrosses indulge in a pre-nuptial display which may vary slightly between species, but of which the basic components are very similar. During these displays the tail is fanned, the wings are opened, the head is outstretched, and the tip of the bill is then buried between the scapulars. These ritualized movements are carried out to the accompaniment of sundry braying and gurgling sounds, except for the Sooty Albatrosses which have a shrill, fluty scream. Whilst most albatrosses basically stand facing each other during display, the Wandering Albatross take things a stage further and step and dance around each other, and in addition the great wings are dramatically outspread. Sometimes several males may gather round a single female and indulge in a communal display. The Light-mantled Sooty Albatross (and possibly also the Sooty) diverges from the pattern just described, and has a synchronous aerial display. Once the nest has been built, or the egg laid, display becomes much less frequent and may not occur at all.

The displays of the albatrosses have rarely failed to fascinate those who have had the good fortune to observe them at first hand, so this brief account of antarctic albatrosses can fittingly be concluded by quoting an account of the display of the Wandering Albatross on Marion Island as observed by the naturalist H.N. Moseley on 26 December 1873, and given in his book *Notes by a Naturalist Made During the Voyage of HMS Challenger* (1879):

> It is amusing to watch the process of courtship. The male standing by the female on the nest raises his wings, spreads his tail and elevates it, throws up his head with the bill in the air, or stretches it straight out forwards as far as he can, and then utters a curious cry, like the Mollymauks, but in a much lower key, as would be expected from his larger larynx. Whilst uttering the cry, the bird sways his neck up and down. The female responds with a similar note, and they bring the tips of their bills lovingly together. This sort of thing goes on for half an hour or so at a time. No doubt the birds consider that they are singing.

This picture shows part of the mutual preening ceremony of the Black-browed Albatross.

Other Birds

This Antarctic Petrel landed on the deck of the Lindblad Explorer *during a storm.*

A Southern Giant Petrel in gliding flight showing the underwing pattern.

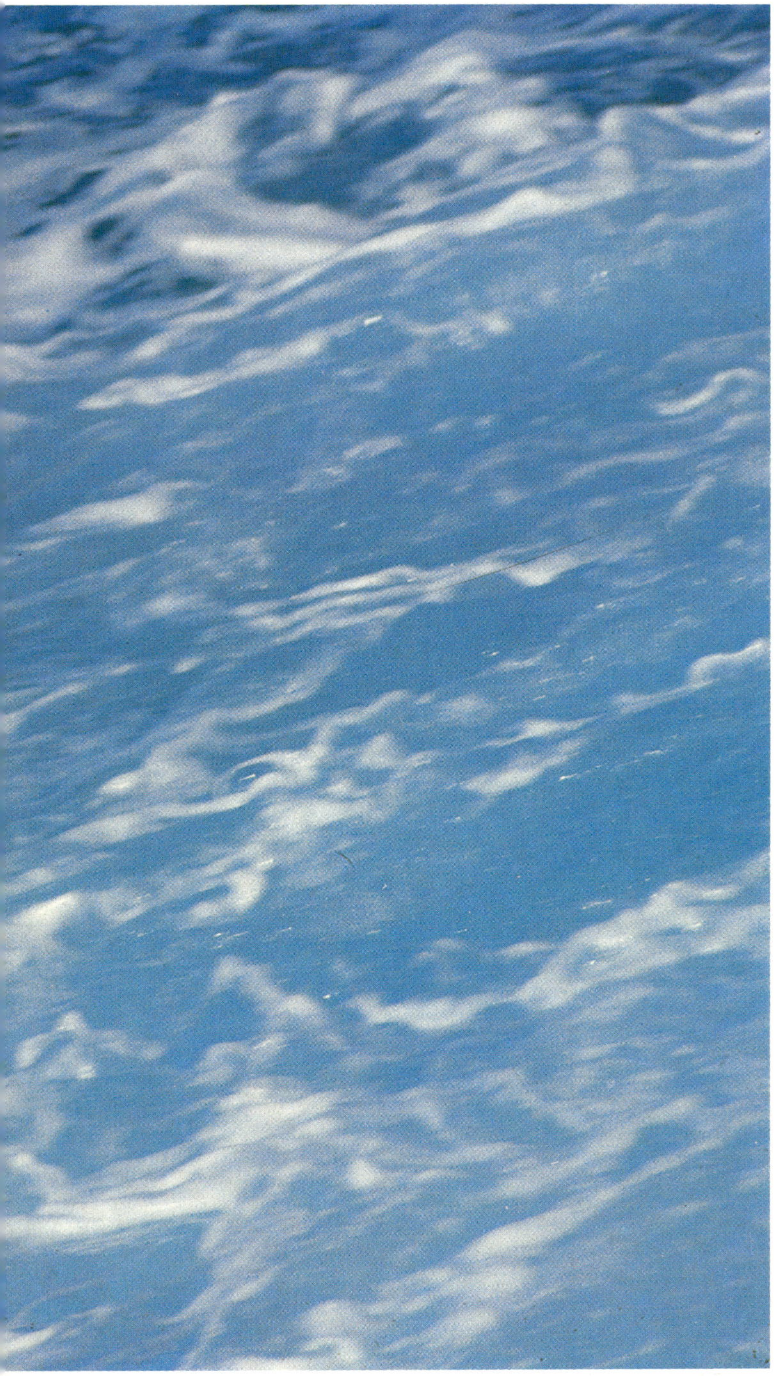

Previous chapters have dealt at some length with the albatrosses and penguins, both large and conspicuous species of seabird. In this chapter we shall look at some of the other birds of the Antarctic, most of which are also seabirds. Only a selection can be discussed, and so non-breeding species will be excluded. Neither can space be devoted to species such as the Sooty Shearwater *Puffinus griseus* which only just qualifies as an antarctic breeding bird by virtue of the fact that it nests on Macquarie Island. Readers will have by now appreciated that seabirds are very abundant and quite important members of the Antarctic ecosystem. It is unlikely that it will ever be possible to arrive at an accurate estimate of the total seabird populations in the area. If counting penguins is difficult, then trying to census the vast numbers of burrow-nesting petrels verges on the impossible. Nevertheless, scientists have attempted by means of very empirical calculations to arrive at some sort of estimate of the total number of seabirds in the Antarctic, and the figure they have arrived at is a minimum of 188 million birds. Even though this figure is obviously subject to a great margin of error, it does give some idea of the level of numbers involved.

Next in importance after the penguins are the petrels of the families Procellariidae, Pelecanoididae and Hydrobatidae, of which there are a couple of dozen species comprising, in numerical terms, over 27 per cent of the total seabird population. With a few exceptions we do not really know a great deal about them. They range in size from very small to quite large and their prey is diverse, running the gamut from minute zooplankton to the larger fishes, squids and carrion.

The numbers of petrels are enormous as anyone who has been on a ship in Antarctic waters will know, but estimating them is complicated by the fact that they come to land only after sundown and, in the case of the burrow-nesting species, one burrow entrance may lead to many nesting chambers of different species. The Cape Pigeon, which is so abundant in subantarctic waters, is an addicted ship follower, and has always been the best-known seabird to sailors in the Southern Ocean. In its shape and manner of flight it is exactly like a scaled-down version of the Southern Fulmar. It reaches the northern limit of its breeding range in the Antarctic at South Georgia. In contrast, the Grey-backed Storm Petrel *Garrodia nereis*, which seems to be uncommon throughout its breeding range, reaches its southern limit at South Georgia. Another not uncommon species of these cold southern waters is the Blue Petrel *Halobaena caerulea* which may often be spotted flitting about in large flocks of prions, and has a wide oceanic distribution. Little has been published about its feeding habits

Above *A Cape Pigeon in flight showing the plumage pattern of the upper parts. This petrel is abundant in subantarctic waters.*

Below *Slender-billed Prion in characteristic flight.*

at sea, but it has been observed to dive and remain submerged for up to six seconds. There is evidence suggesting that it commonly falls prey to skuas, since 173 corpses were collected from skua middens at Pearson Point, on Bird Island, South Georgia, between December 1971 and February 1972. On Annenkov Island about 40 per cent of several hundred skua-killed corpses examined in 1972-3 were Blue Petrels.

Two species, the Antarctic Petrel and the Snow Petrel, rarely get far beyond easy flying distance of the pack-ice, and the two species frequently associate together. Sometimes sizeable flocks may be seen swarming around the icebergs like gnats, and large flocks may often be encountered resting on floes and icebergs. Both species nest deep within the Antarctic Continent, laying their eggs in crevices of the rocks on steep scree slopes, or on exposed cliff ledges. They are able to find nest sites on the exposed mountain peaks or nunataks that project through the ice of the interior plateau. By the early 1970s only nine breeding colonies of the Antarctic Petrel were known on the continent, but others have since been discovered, particularly in the Russian and Australian sectors, although the details do not seem to have been published. Large numbers of Antarctic Petrels, and small numbers of Snow Petrels, nest in the Theron Mountains (79°S, 28°W) some 250 kilometres inland from the Weddell Sea. The Antarctic Petrel colony on Mount Faraway in the Theron Mountains was at one time the most southerly breeding site known for any species, but there are now unpublished data on localities even further south. So far as is known, the largest colony of this species is the one discovered in January 1960 by the Soviet Antarctic Expedition at Muhlig-Hofmannfjella in Queen Maud Land; it was estimated to contain about one million birds. Studies of Snow Petrels in the

Pointe Geologie Archipelago showed that the annual mortality of eggs and chicks was no less than 52.7 per cent, a sure indication of the severity of their environment.

Wilson's Storm Petrel, like the two species just mentioned, belongs to that small and elite group of species that nest on the Antarctic Continent itself. It is known to breed in crevices in cliffs and scree slopes all around the edge of the continent, and it may be that at some time in the future it will be discovered nesting, like the Antarctic and Snow Petrels, in the high mountains of the interior.

The various species of petrel discussed above do not present any great problems as regards identification in the field, but the same cannot be said for the prions or whalebirds (*Pachyptila* species). These small bluish-grey petrels with white underparts are specially adapted to feed directly upon the small zooplankton found near the surface of the sea. This they do by straining the water through lamellae fringing the bill. The five antarctic breeding species are so similar on the wing that identification is more often than not quite impossible for the observer standing on the heaving deck of a ship. They have a highly erratic and buoyant flight and commonly feed in dense flocks. Only the Antarctic Prion nests south of latitude 60°S, and it is thought that it may now be extinct as a breeding species on the Antarctic Continent.

Anyone seeing these birds in the Southern Ocean, apparently in their millions, could be forgiven for thinking that they must be the most abundant birds in the world. The American ornithologist Robert Cushman Murphy wrote of them:

> an observer who has seen the flights of such southern forms as *Pachyptila* (the prion genus) filling the air like the flakes of

Above *Fairy Prion. Note the large webbed foot.*

Overleaf *Giant Petrels are inveterate scavengers. This group are fighting over food on the sea.*

Below *Head of Fairy Prion showing tube nose.*

Left *A Southern Giant Petrel incubating. Note the detail of the bill structure.*

Below *Part of an aggressive display by an Antarctic Skua. Note chick below the wing.*

A young Southern Giant Petrel expresses its disapproval.

> a snowstorm, and stretching in all directions towards the circle of the horizon from daybreak until dark . . . might be pardoned for extravagant assertions.

We will stay with Murphy for the moment, for it would be difficult to find a more poetical description of the beautiful dancing flight of these birds:

> The erratic gliding of these petrels is the most wild and airy type of flight among all birds: it might best be described by the term 'oestrelatous' – goaded on by a gadfly. When the air is filled with a flock of whale-birds careening in the breeze, rising, falling, volplaning, twisting, sideslipping above the sea, now flashing their white breasts, now turning their almost invisible backs – they resemble the motes in a windy sunbeam.

The Fairy Prion *Pachyptila turtur*, the only prion species illustrated in this book, breeds on Marion and Prince Edward Islands and Îles Crozet. On the former island group, Salvin's Prion *Pachyptila salvini* nests in burrows on cinder slopes up to 600 metres above sea level; it is probably the most numerous petrel in the islands. Similarly, the Antarctic Prion is probably the commonest breeding species on South Georgia. Prions are frequent victims of marauding skuas and some years ago, on Bird Island, South Georgia, it was estimated that 10,000 prions were taken by skuas during each breeding season. In order to sustain this level of predation their numbers must obviously be very high.

In contrast to the prions, even the novice bird-watcher should have little trouble in recognizing the two species of giant petrel – the Northern Giant Petrel *Macronectes halli* and the Southern Giant Petrel. They are quite unmistakably fulmarine petrels, closely related to the Southern Fulmar, but in size equal to the smaller species of albatross (the mollymawks). A distinctive characteristic is that the quite enormous bill seems oddly out of proportion in relation to the relatively small head. The breeding range of the Southern Giant Petrel is from the Antarctic Continent north to islands near the Antarctic Convergence, while the Northern Giant Petrel nests on islands close to the convergence and extends north to the Subtropical Convergence. At four localities – Macquarie Island, South Georgia, Marion and Prince Edward Islands, and Îles Crozet – their ranges overlap and both species nest side by side as it were (sympatrically, in scientific jargon) without effective hybridization. As it so happens, mixed pairs have been noted on more than one occasion but the eggs have failed to hatch, so 'effective' is clearly the key word here. There are differences between the two species in the choice of nest site, nesting habits, and in the timing of breeding. The southern species completes laying three weeks before the northern species begins. The Southern Giant Petrel has both dark and light plumage phases, but no white phase is known in the northern species. Like the Wandering Albatross, it is probable that the early years of life of these petrels involves circumpolar wanderings in the west-wind drift.

The giant petrels were well known to the sealers since crowds of them would invariably assemble, like vultures, when seal carcases were being stripped of skin and blubber on the beach. The men referred to them variously as 'Breakbones', 'Nelly' or 'Stinker'. Like the fulmars and other petrels, the giant petrels have an effective means of protection when provoked, something which clearly irritated the naturalist H.N. Moseley when he was on Marion Island in December 1873:

> I found a young bird, I think the young of the Giant Petrel, in a nest scarcely raised from the ground; the young bird vomited up the contents of its stomach and gush after gush of red oily fluid at me as I stirred it up with a stick. All the petrels vomit oil in this way, and the white ones thus are apt to spoil themselves for stuffing in a most provoking way, before one can get their mouths and nostrils stuffed with cotton wool.

The situation with these two species of giant petrel on South Georgia is quite interesting. The present breeding population of the southern species there is 20,000–30,000 pairs, and of the northern species 5000–15,000 pairs. The recent population explosion of the Antarctic Fur Seal on Bird Island has had some interesting effects on the population of the latter species. Between 1973-4 and 1978-9 the Northern Giant Petrel population on Bird Island increased from less than 500 pairs to 1100 pairs, while the population of the Southern Giant Petrel showed almost no increase. The increase in the population of the northern species coincided with that of the fur seal population because there was an increase in the availability of placentae and pup carcases in December and January for food. Bearing in mind the difference in the timing of breeding of the two petrels, it is obvious that this increased food supply is available during the chick-rearing period only to the Northern Giant Petrel, so it is not unreasonable to assume that it has been a significant factor in the population increase of that species.

Whether or not the Brown Skua and the South Polar Skua along with the Great Skua of the northern hemisphere, are all

This weka, photographed on Macquarie Island, is an introduced species from New Zealand.

Above *A South Polar Skua in flight.*

Overleaf *A King Cormorant incubating on its bulky nest. This species has a restricted distribution on islands of the Subantarctic.*

Above *Male and female Antarctic Cormorants. This is the only species of this family to breed in the Maritime Antarctic.*

Above right *A pair of Antarctic Cormorants with a young bird.*

Below right *A breeding pair of King Cormorants indulge in a bout of mutual preening.*

Overleaf *An Antarctic Cormorant surveys its snowy domain.*

races of one species is a question that is certainly complex. In some areas where both species breed mixed pairs can be found. On King George Island in the South Shetlands, for example, studies in one particular area in 1977-8 revealed the presence of 16 pairs of Brown Skuas, eight of South Polar Skua, and four mixed pairs. Similarly, on Anvers Island near the Antarctic Peninsula, earlier studies of Antarctic Terns, on whose nests skuas are principal predators, showed that there was a complex hybrid population of skuas.

Skua populations are increasing in some areas, although not necessarily for the same reasons in each case. At Pointe Geologie, Adelie Land, on the Antarctic Continent for example, the increase of the South Polar Skua population from 80–90 in 1965-6 to 320 in 1976-7, was considered to be caused by the establishment of a garbage dump. Skuas are now found in summer at every human settlement along the rim of the continent. No sooner had man arrived in the Antarctic than the skuas adapted to his presence, with considerable profit to themselves. At South Georgia the activities of the expanded fur seal population in flattening extensive areas of tall tussock grass on Bird Island, may be partly responsible for the increase in the Brown Skua population from 175 pairs in 1958-9, to about 350 pairs in 1976-7, simply by increasing the area of suitable breeding habitat.

The skuas are, of course, traditionally associated with the penguin rookeries although, apart from human settlements, there are other sources of food, including the afterbirth when the seals give birth around the coasts. They are opportunistic feeders and have even been seen to take milk from lactating Southern Elephant Seals alongside the pups, and after the pups have been fed.

Left *An adult Wattled Sheathbill looks down at its young at the base of the rock fissure.*

Above *Sooty Shearwater.*

Every penguin rookery will have its attendant skuas, hovering and wheeling overhead, perhaps diving down to reappear a few seconds later with an egg in the bill. The eggs are normally carried away from the vicinity of the rookery to be consumed at leisure. Newly hatched penguin chicks are not often taken as they are usually brooded by the parents. Once the brooding period is finished and the chicks are left alone they tend to fall victim to the skuas, although they are usually safe enough so long as they remain in the crèches. The close relationship between skuas and penguin rookeries has been pretty well studied. Work carried out in the 1977-8 breeding season on King George Island, South Shetlands, showed that skua territories contained from 90 to 2011 penguin nests. 'Optimal' skua territories contained from 766 to 2011 penguin nests, and 'sub-optimal' territories from 90 to 260 nests. The skuas defended these territories against intruders such as other skuas, Dominican Gulls and Wattled Sheathbills. What these studies showed was that skuas that defended feeding territories preyed exclusively on penguin eggs and chicks once these became available in late October. Those skua pairs defending optimal territories fledged significantly more chicks than all other skuas.

The most southerly breeding record for the South Polar Skua appears to be at Mount Faraway in the Theron Mountains where in January 1967 six pairs were found defending territories. The occasional skua is likely to be seen anywhere in the interior of the Antarctic Continent, having been recorded not only at the South Pole itself, but also at the Soviet Union's Vostock Station which lies at an altitude of 3488 metres above sea level.

Among the other breeding seabirds of the Antarctic are the shags or cormorants. They differ from the penguins and petrels in that adults may be present in the vicinity of their breeding sites throughout the year, and are rarely seen far from land or the ice edge. They feed in inshore waters, and the Antarctic Cormorant or Blue-eyed Shag is known to be capable of diving to depths of at least 25 metres. The Antarctic Cormorant is the only species of this family to breed in the Maritime Antarctic, where its colonies are often associated with the rookeries of Chinstrap and Gentoo Penguins. Individuals returning with food are very liable to be relieved of this by the piratical skuas, and whilst out at sea cormorants are quite likely to be attacked by Leopard Seals. The closely related King Cormorant *Phalacrocorax a.albiventer* and the Kerguelen Cormorant *Phalacrocorax albiventer verrucosus* have a rather limited distribution on islands of the Subantarctic. Their habits are similar to those of the Antarctic Cormorant but they construct much bulkier nests, and the breeding cycle is more prolonged.

The Dominican or Kelp Gull is the only gull to nest in the Antarctic where it has a wide distribution from the islands of the Subantarctic south to the Antarctic Peninsula. In the more southerly breeding locations bad weather can cause extensive failure of early nests. Considerable mortality of eggs and young can result from the depredations of skuas and sheathbills. Their distribution and numbers during the winter are influenced by weather and the availability of open water and food. This gull feeds primarily on limpets, but also scavenges and will take advantage of carrion.

Finally, mention must be made of those curious and somewhat ugly pigeon-like birds, the sheathbills. They are not seabirds but land birds, and the Wattled Sheathbill is in fact the only land bird to breed in the Maritime Antarctic. The Lesser Sheathbill *Chionis minor* is confined to four island groups in the Subantarctic. Although capable of strong flight, the sheathbills spend most of their time running and hopping over the beaches or rocks. They are voracious scavengers of eggs and almost anything else edible, and are typical hangers-on at penguin and cormorant colonies. They have even been seen to interrupt feeding penguins in order to rob the chick of its meal of regurgitated krill or fish. In the winter they habitually visit camp refuse dumps. Sheathbills are incurably inquisitive, an aspect of their behaviour well described by H.N. Moseley who encountered them on Marion Island in December 1873:

> Living also about the rookery (of King Penguins) was a flock of about thirty Sheath-bills (*Chionis minor*). The instant they saw us approaching they came running in a body over the floor of the rookery in the utmost excitement of curiosity, and came right up within reach of our sticks, uttering a 'Cluck, cluck' which with them is a sort of half-inquisitive, half-defiant note. We knocked over several with big stones and our sticks; but the remainder did not in the least become alarmed.

Tameness and curiosity are a feature not only of the sheathbills, but also of many other birds and animals of the Antarctic, and a similar situation often prevails in the Arctic. Fortunately, visitors to the Antarctic today are less lethally armed with cameras and field glasses, rather than sticks and stones.

A group of Southern Black-backed, Kelp or Dominican Gulls on an iceberg.

A Southern Fulmar in flight showing the underwing pattern. The flight alternates between fast wingbeats and long glides.

Seals

It was 200 years ago that Captain James Cook circumnavigated the Antarctic, and what he subsequently wrote about the whales and seals was of enormous interest to the whaling and sealing industry of the northern hemisphere where catches were already declining. On hearing of Cook's report, sealers of various nationalities rushed southwards to begin yet another orgy of slaughter. Just precisely what impact these men had on the seal populations during the course of the next 50 years we shall never know, but what we do know does not make pretty reading; virtually nothing was left that could be easily and profitably slaughtered. The species worst affected were the fur seals (for skins) and the elephant seals (for oil), and the populations of these have shown substantial recovery only in the last 30 years or so. The later sealers always had a major interest in blubber and oil, and the industry based on these largely superseded that based on fur. Other antarctic seals – Crabeater, Leopard, Ross and Weddell – were mostly ignored by sealers as unsuitable to the trade, although the Crabeater Seal was exploited to some extent. Because of the circumpolar distribution of these species in the pack-ice zone, very little was known of their life history.

Although depressing, it is certainly illuminating to look at some of the gory details of the exploitation of the seals. No sooner had Cook's account of his discoveries seen the light of day than British sealers, closely followed by Americans, hurried to South Georgia. By 1791 there were at least 102 ships exploiting the stocks of Antarctic Fur Seals and Southern Elephant Seals, and sealing there reached a peak about 1880-1. James Weddell (himself a sealer) estimated that by 1882, 1,200,000 skins had been taken and that the fur seal was to all intents and purposes extinct on the island.

The same bloody story was repeated at many other islands in the Southern Ocean. Macquarie Island was first sighted by the brig *Perseverance* in 1810, and the first vessels to drop anchor there found the beaches crowded with seals. The sealers wasted no time and around 90,000 were butchered in the first season, 60,000 in the second, and 20,000 in the third. By the following season only a pitiful handful of seals were left. Further south, the South Shetland Islands were discovered by William Smith in 1819. The focus of sealing swiftly shifted south and in the 1820-1 season sealers from no less than 47 American and British ships were working the beaches of these islands. During this peak year about 250,000 fur seals were taken, and many thousands more killed and lost. Considerable numbers of Southern Elephant Seals were also boiled down for oil.

How many seals are there at present in the Antarctic? As is the case with most such questions, the answer is not easy to

Right *A group of Crabeater Seals on the pack-ice.*

Below *The Crabeater is the most abundant of the Antarctic seals and its population may be as high as 16 million.*

Crabeater Seals with attendant Dominican or Kelp Gulls.

provide. However, in 1977 Dr Richard M. Laws, Director of the British Antarctic Survey, made an attempt to produce a total population estimate. The figures he arrived at were: Crabeater Seal 14.86 million, Weddell Seal 730,000, Leopard Seal 220,000, Ross Seal 220,000, Southern Elephant Seal 600,000 and Antarctic Fur Seal 200,000. This gives a total population of approximately 17 million seals which, incidentally, were estimated to consume about 80 million metric tons of krill, squid, and fish each year. Obviously these estimates are subject to large possible errors, but at least they provide a reasonable guide. The latest available data suggest that the population of the Crabeater Seal may now be around 16 million. The Antarctic Fur Seal population on South Georgia alone is now estimated at 500,000, and the beaches there are now so crowded that accurate estimates are very difficult. In addition, the Southern Elephant Seal population has probably increased and may now be nearer 700,000. Taking these increases into account gives us a total population of seals in the Antarctic of something in excess of 18 million.

It is obvious from the figures given above that the Crabeater is the most abundant of the antarctic seals, and it is in fact the most numerous seal in the world. Although it has a circumpolar distribution in the pack-ice, it is restricted to the fringes; research has shown that it is consistently more abundant in areas dominated by cake and brash ice, and this distribution is related to the availability of food since the krill exploited by Crabeater Seals has been found to be most prolific in proximity to the pack-ice edge. The dentition of this seal is highly specialized to act as a sieve to strain krill.

Crabeater Seals probably also utilize the pack-ice to avoid predators. When young they are subject to heavy predation by Leopard Seals and Killer Whales *Orcinus orca*. Very nearly all Crabeater Seals more than a year old bear characteristic scars. Since a large proportion of adults also show these scars, the inference is that from around one year of age onwards they have a high escape rate from predators. The danger is in the water; once on an ice floe the Crabeater is pretty safe. There are very few records indeed of a Leopard Seal attacking prey on an ice floe. Not all that much is known about the breeding habits of this seal, since pupping apparently occurs in areas of heavy pack-ice during the austral spring.

The Leopard Seal is also primarily a resident at the fringes of the pack-ice zone. It does, however, have a wider range than the other antarctic seals (probably because of its varied feeding habits), and is found on the continental coasts, fast-ice, pack-ice, and to some extent on the subantarctic islands. Studies at Palmer Station, Antarctica, have produced results supporting a theory that this species is at least partially migratory, summering in the antarctic pack-ice and wintering in the subantarctic islands. The Leopard Seal is a predator, and it is indeed the only seal that regularly feeds on warm-blooded animals. It has a graceful streamlined body, and a disproportionately large reptile-like head. When disturbed it raises the head alertly thereby exposing the white and black blotched throat.

The Leopard Seal is best described as an opportunistic feeder. Much has been made about its habit of preying on penguins, patrolling the shoreline off the rookeries where it stays just outside the breakers and faces out to sea. When a group of returning penguins is spotted the seal sinks beneath the surface in order to surprise its prey. It has been estimated that at the Cape Crozier Adelie Penguin rookery, Leopard Seals account for some five per cent of the entire breeding population of the penguins each season. Nevertheless, it is considered that the Leopard Seal does not in fact control penguin numbers through predation. Evil looking though the Leopard Seal is, when hauled out of the water penguins will walk unconcernedly past within easy reach without appearing in the least troubled. Like the Crabeater, the Leopard Seal's dentition is adapted to sieve krill, and krill frequently predominates in faecal samples. Studies in the Pacific sector of the Southern Ocean have shown that Leopard Seal densities were associated with Crabeater Seal densities more often than with other seals or penguins. It could be argued that this correlation is the result of a common food source (krill), but there is little doubt that Leopard Seals, particularly adults, eat other seals as well as krill, and seal remains are frequent in the stomachs of Leopard Seals that have been captured on the pack-ice.

The Weddell Seal is the only higher animal (excluding man) that is not a seasonal visitor to the Antarctic. It has evolved to fill a very specific niche in the antarctic ecosystem – the inshore fast-ice zone. It survives the winter by remaining in the warmer water below the sea-ice, warmer in this instance being entirely relative to the bitter cold above the ice. Being an air-breather, the only way this seal can survive from autumn to spring when the sea is frozen and the air temperatures too low for it to emerge safely, is by using breaks in the ice produced by the wind, tide and currents, or by the maintenance of breathing holes in the ice. This it does by using its procumbent incisors and canines to rasp away at the ice. Failure to do this means death by suffocation. There is a fairly high adult mortality in this species, and it has been suggested that a major reason for this is a higher rate of tooth wear, inevitably reaching the stage where the animal cannot maintain its breathing hole. In the McMurdo Sound area of Antarctica, for example, it has been estimated that from the age of eight years onwards, the average annual mortality rate is just over 27 per cent.

The Weddell Seal tends to have a more southerly distribution than any of the other antarctic seals. Despite the obvious hazards of living in the fast-ice zone, its ability to utilize a habitat unavailable to other seals has its compensations. It is, for example, protected from predation by Killer Whales, and it is able to exploit inshore prey which is primarily fish; but it also takes some benthic animals and squid elsewhere. Its distribution is circumpolar and it is relatively sedentary, although it has been reported as a straggler from most of the subantarctic islands. There is nothing random about the distribution of Weddell Seals on the ice. A well-known feature of this species is its high fidelity to pupping areas, and well-defined rookeries are located along perennial pressure cracks in the ice. Typically, these are composed of pregnant females, females with pups and a few mature males. Male territories are spaced under the tide cracks, and copulation takes place under water. On the periphery of the rookery will be found immature seals, non-pregnant females and other males. As the season progresses, so the numbers at the rookeries decrease as the animals disperse among the expanding tide cracks in the ice. Notably absent from the pupping areas are animals of the younger age groups, that is to say up to about six years old, and it is likely that these are away in the pack-ice.

Because the Weddell Seal inhabits the fast-ice zone it can be studied by scientists at antarctic research stations, and as a result a good deal is now known about it. Even so, there still remain a number of unanswered questions concerning this unique animal. Studies in its natural environment have produced some fascinating results. It is now known, for example, that it can spend an hour in the water without surfacing for air. In

The predatory Leopard Seal has a graceful, streamlined body and a disproportionately large head.

A Leopard Seal on the ice presents no threat, but in the water it preys on some other seals and penguins.

Above *The Weddell Seal is well adapted to the severe conditions in the fast-ice zone in which it lives.*

Right *A Weddell Seal looks curiously at the photographer but shows no fear.*

pursuit of food it undertakes deep dives to depths of 300–400 metres, and can go as deep as 600 metres. Although these deep dives are usually of short duration – seldom more than 15 minutes – they may be continued for several hours with only brief rests in between. It is also known that the Weddell Seal can descend (and rise) at a rate of 120 metres per minute, yet suffers no ill effects from the increase in pressure, neither does it suffer from decompression ('the bends') on its return to the surface. How is this achieved? We get a clue from the animal's anatomy, which indicates that it copes with pressure increase by yielding to it. For example, its main body cavities are so constructed that they readily compress, and it has a very flexible rib cage. In addition the diaphragm underlying the lungs is at an oblique angle to the chest cavity. It seems probable that external pressure can cause complete collapse of the lungs, pushing gases into the bronchi and trachea, thus permitting the seal to avoid lethal differences between external and internal pressure. Lung collapse at depth means an absence of nitrogen in the lungs, with no absorption of that gas into the bloodstream, and thus no 'bends'. Indeed the Weddell Seal is a very remarkable animal.

The Ross Seal has an irregular circumpolar distribution in the pack-ice (primarily consolidated pack-ice) surrounding the Antarctic Continent. Very little is known about it. There is, for instance, almost no information about its activity patterns, breeding habits, behaviour, or ecology – none of which is surprising in view of its habitat. In all the years of antarctic exploration, less than 200 sightings had been made prior to 1972. Most sightings of this species are of solitary individuals, although groups of 20 or more may occasionally be seen. It does not seem to collect in large breeding groups, and pupping is

thought to occur in the heavy pack-ice in November and December.

Quite why Ross Seal densities in the pack-ice should be as low as they seem to be is not known. It is certainly not due to man-induced mortality, so perhaps food resources (it feeds primarily on cephalopods) are a limiting factor, and possibly there is some degree of interspecific competition with Crabeater and Weddell Seals. When disturbed, the Ross Seal adopts a characteristic posture with the head pointed straight up and the throat inflated. This is the singing posture, and a unique feature of this seal is its bird-like vocalization which is produced by grotesquely inflating the throat and slowly exhaling. There is no mistaking the Ross Seal, for apart from the features just described it is short and fat with a rather frog-like head, large eye orbits and broad flared nostrils, which can be closed so tightly that they become almost invisible.

The Southern Elephant Seal breeds on nearly all the subantarctic islands around the Antarctic Convergence. The three main breeding populations are on South Georgia, Macquarie Island, and Îles Kerguelen respectively, which together probably hold well in excess of half a million animals. With the recovery from earlier exploitation, the elephant seal has been extending its range southwards in recent decades, and small breeding colonies have been established in the South Orkneys and South Shetland Islands. At Signy Island in the South Orkneys natality and pup survival has varied in relation to ice conditions. Some of the freshwater lakes on Signy Island receive nutrients from seal and seabird colonies, and seal stocks are increasing to the point where their effluents are threatening some ponds with eutrophication. The southernmost regular breeding site for the elephant seal appears to be Stranger Point on King George Island (62°16′S, 58°37′W) in the South Shetlands. In October 1976 a female with a pup, together with a group of a dozen subadult males and two adults of undetermined sex, were seen more than 240 kilometres south in the Gerlache Straits at 64°50′S, 62°40′W. There is no other evidence of elephant seals breeding in this area. Large concentrations of non-breeding elephant seals are found in the South Shetland Islands, and as many as 25,000 have been counted.

The elephant seal is the largest of the antarctic seals, and a fully grown adult bull may be six metres long and weigh around four tonnes in full fat; cows are usually much shorter and lighter. On returning to the breeding beaches after wintering at sea, mature bulls stake out a territory, and the returning cows collect around them in huge harems. Needless to say, the formation and maintenance of a harem, and the repelling of rivals, involves the bulls in pretty continuous violent displays of aggression and fierce fights. It is usually a case of the biggest man wins, so it is not surprising to find that most breeding bulls are normally at least 12 years old. The aggressive displays are accompanied by frightful deafening roars produced from the great inflatable nostrils which hang down over the mouth and so form a resonating chamber. Throughout the breeding season, which varies with locality, the bulls fast and live off their enormous reserves of blubber.

Finally, there are the fur seals of which we are concerned with three species – the New Zealand Fur Seal *Arctocephalus forsteri*, the Kerguelen Fur Seal and the Antarctic Fur Seal. Within the Antarctic the New Zealand Fur Seal breeds only on Macquarie Island, and we shall not discuss it further. The remaining two are primarily species of the Maritime and Subantarctic. The food habits of these seals in the vicinity of their breeding grounds show distinct differences. The Kerguelen

Fur Seal is a more general feeder, taking mainly squid, Nototheniid fish and Euphausid crustacea. The staple food of the Antarctic Fur Seal in South Georgia waters is the krill *Euphausia superba*. This swarming krill is entirely confined to waters south of the Antarctic Convergence where it is extremely abundant. It must be emphasized that these remarks apply only to the breeding season, since nothing is known of the winter distribution of these fur seals, nor of their food habits at that time.

The Kerguelen Fur Seal now breeds only on Marion Island where, incidentally, the Antarctic Fur Seal also breeds. It used to be thought that the Antarctic Convergence separated these two species, but it is now known that this is not the case. An Antarctic Fur Seal tagged on South Georgia has been recovered in South America, and in recent years there have been several

Above *Southern Elephant Seals lying side by side amongst seaweed on the shore of a subantarctic island.*

Overleaf *A male Southern Elephant Seal holds forth in no uncertain fashion.*

records of the Kerguelen Fur Seal at South Georgia several hundred kilometres south of the Antarctic Convergence. A young Kerguelen Fur Seal has also been seen at Macquarie Island, over 5000 kilometres from the species' nearest-known breeding station. The Antarctic Fur Seal now breeds also on Îles Kerguelen, Heard and MacDonald Islands, Bouvet Øya, South Georgia, and the South Sandwich, South Orkney and South Shetland Islands. However, the main breeding colonies are those on South Georgia. Research there showed that the mean annual rate of increase from 1958-9 to 1972-3 was 16.8 per cent overall.

During the winter from about May to October, both sexes are somewhere out at sea. The adult bulls begin to come ashore towards the end of October some two or three weeks before the cows haul out to give birth to the pups. Over 90 per cent of the pups are born within a three-week period from mid-November, and over a three-year study period the date by which 50 per cent of births had occurred varied by not more than one day. The growth of the pup during the 110–115 days suckling period is quite phenomenal, and has been estimated at 98 grams per day for males and 84 grams per day for females. There is no other species of fur seal that grows so quickly at this time.

When the first bulls come ashore in late October there is plenty of room on the beaches, so very little fighting takes place. This situation changes very quickly, and the later arrivals have to fight to establish a territory. It was found that as a result of such fights the mean territory size decreased by about one third between mid-November and December. The first bulls ashore usually establish their territories at the water's edge, while later arrivals form territories inland of these. If a bull is charging inland through a series of territories, or towards the sea, then each bull whose territory he passes through, along with neighbouring bulls too, will chase him, but only as far as the edge of their respective territories where pursuit ceases. Territorial boundaries are vigorously defended by displaying with neighbours. These boundary displays are ritualized threat displays and their form is quite stereotyped: the bulls meet with heads pointed high then shift to a 'facing away' posture. They may return repeatedly to the head-high position, sometimes with an 'oblique stare', as many as four or five times in a single display, but once or twice is more usual. It is in these territories that for a period of up to two months that the cows, in small harems of four or five, give birth to their pups, and mate. Territorial bulls start to abandon their territories in late December, by which time the majority of the cows are mated and are making feeding trips to sea, then returning to feed their pups. The pups are like small woolly dogs, yapping noisily at one another, and rolling and fighting in the often glutinous mud of the breeding colonies. Fur seal colonies with their noise, smells and sundry disputes, are not exactly the most attractive of places in which to spend much time, but there is no denying their fascination.

A Southern Elephant Seal in restful mood.

Wildlife Photography in the Antarctic

The Antarctic is a very photogenic place and photography there can be both exciting and rewarding, and the results quite lovely. Nevertheless, the area presents problems to the photographer very different from those experienced in warmer parts of the world. The birds and mammals, whose acquaintance with man is relatively recent, are tame and confiding. There is no need for high-power telephoto lenses, and most of the photographs reproduced here were taken with lenses ranging from 75–150mm. For most of the flight shots, taken from the stern of the ship, Eric Hosking used an f/2.8–180mm Zuiko lens, and whenever the light was good enough this was stopped down to f/5.6. If the bird was flying towards the camera an exposure of about 1/250th second was given, but when flying across the line of vision, especially with a following wind, 1/1000th second was scarcely sufficient to stop movement.

With modern automatic cameras, such as the Olympus OM-2n, the photographer sets the lens at the aperture required and the necessary exposure is calculated automatically. However, it is essential not to rely exclusively on the camera for it is an excellent servant but a very poor master. One moment a bird may be flying against a bright sky, the next against the dark blue of the sea. The camera gets confused by the intense brightness of the sky and darkness of the sea. Uncorrected, it will in the first case give an over-exposed picture, and in the second an under-exposed picture. In conditions like this the trick is to bracket exposures and take several shots using the over-ride on the camera. This can often be done even without taking the camera down from the eye. When the bird is flying against brilliant, white cumulus cloud, increase the shutter speed by the equivalent of perhaps two stops, or when it is skimming close to the deep blue of the sea, *decrease* it by as much as two stops. The same principle applies just as much when on land. Penguins on the snow, with the sun shining on them, give such a brilliant scene that one may be unwilling to believe an exposure of 1/1000th second at f/11 on Kodachrome 64. With elephant seals lying in a heap on black volcanic ash with a heavy sea mist as well, a long exposure of 1/10th second at f/3.5 on Ektachrome 200, for example, will be necessary.

Photography is full of compromises. A slow film, such as Kodachrome 25, gives much better colour balance and definition than Ektachrome 400, so obviously use the former whenever possible. But there are frequently times when lighting conditions and the speed that a bird is flying make it imperative to use the faster film in order to obtain any result at all. It is best to use the shortest possible focal length lens that the situation allows, and the camera will be much easier to hold steady. However, if

Overleaf *A Black-browed Albatross in flight showing the pattern of the upper surface of the wing.*

Below *Sooty Shearwaters taking off from the sea. Within the Antarctic this shearwater breeds only on Macquarie Island.*

you wish to obtain a close-up of the head of a King Penguin you will be within centimetres of it if a short focus lens is used and it may become nervous, so it is best to use a 100mm or a 135mm lens.

There is often a tendency for flash to be used even when there is no need for it, and even in cases where it will be totally ineffective. A group was once observed attempting to photograph the huge caldera of a volcano on Deception Island, obviously without realizing that the light of the flash would not penetrate more than about 10 metres, whereas the far rim of the caldera was over a kilometre away. However, flash can be invaluable for many subjects, not only as the main source of lighting but also to obtain a high-light in the animal's eye. In such a case the light from the flash will carry a considerable distance, being reflected from the bright surface of the eye like a mirror. Flash is also useful for taking photographs of birds in dark places, such as a petrel brooding over eggs near the entrance to its nesting hole.

One of the greatest problems that the photographer has to overcome is camera shake. The best answer is to use a good, heavy, stout tripod, but who really wants to carry one of these around all the time? In any event a tripod is not much use on the deck of a ship with the engine vibrations going through it. There are various gadgets that can be tried, but probably the best under most circumstances is a monopod fitted into a leather flag-holder like those used by scouts. When this is worn across the shoulder the monopod slips into the slot made for the flag mast. If the camera on the monopod is then pulled down firmly it is surprising how steady it becomes, and it is quite feasible to give a 1/60th-second exposure with a 400mm telephoto lens and still avoid camera shake.

It is best when standing on the deck of a ship to relax the knees to avoid as much vibration as possible, and it is also a good idea to use a rope tied round the waist and the ship's rail. This allows one to lean back so that there is no need to worry about maintaining balance on the rolling ship, and one can concentrate on getting photographs of the birds following in the wake. It is a good idea to estimate the distance the bird is away from you, then to set that distance on the lens and wait for the bird to fly into focus, firing the shutter a fraction of a second before it reaches critical definition. This is necessary since there is a delay in the message from the brain reaching the fingers and thence the shutter release. By using this method it will be found that a fairly high percentage of sharp pictures will be obtained, whereas if one tries to keep continual focus on the bird it usually flies through the critical zone before the shutter fires.

Overleaf *Adelie Penguins porpoising in the snow.*

Right *A volcanic caldera on Deception Island in the South Shetlands.*

If the weather becomes very cold, well below freezing, and especially with a bitter wind blowing, some batteries fail to work. It is best in such conditions to keep the camera and motor drive inside the outer clothing until ready for use. The modern automatic camera with its printed circuits is not affected by the cold, although the batteries might be, but mechanical equipment can be affected due to the thickening of the oil which can slow down the shutter and cause over-exposure.

Always keep UV (ultra-violet) filters on all lenses, mainly because they give a better colour rendering, but also to prevent the lens itself from getting scratched; it is far cheaper to replace a filter than a lens! In the normal course of events there will be no need for other types of filter although there are times when they can be useful. For instance, with the sun shining on the snow the whites tend to burn out, even though you compensate for this, and a pale-blue filter helps to add tones. Again, a deep-red filter can, under certain circumstances, give some wonderful sunset effects even if the picture is taken in the middle of the day!

It is a very good idea to make notes of the photographs taken and to number each film accordingly, otherwise when you arrive home sometime later it can be very difficult (even impossible) to remember one place from another. However, because it is often too cold to write, or too inconvenient to get at a notebook and pencil stowed away in a pocket, a tiny electronic notebook is ideal. It is so easy to speak into and to dictate all the relevant information while you are walking along, and when it is really cold it is best to keep moving.

Although not strictly wildlife photography, it is always worthwhile producing a picture story to show to others after the expedition. They will not only want to see shots of the birds and mammals, and maybe the vegetation as well, but will also appreciate getting some idea of the conditions under which you have worked. Therefore, take a flash picture of the interior of your cabin on the ship (a wide-angle lens will be required for this); shots of the ship among the icebergs; shots of yourself and others dressed to face the icy, bitter cold; and of the rubber Zodiac boats used to transport people from ship to shore in the Antarctic, and similar scenes.

Some participants in antarctic cruises may have the extreme good fortune, as did Eric Hosking, to be landed close to Captain Scott's hut at Cape Evans. Here is the setting for what was one of the great milestones in the history of antarctic exploration, and an ideal photographic subject. Entering the hut, as mentioned earlier in this book, can only be described as an uncanny experience. The interior of the hut is just as it was when those brave men left it for the last time. So much so that one feels very vividly that the expedition members had only just left and would be back at any moment. To photograph this successfully is quite a challenge; flash is essential, and so is a range of lenses from 18mm super wide-angle to include as much as possible of the interior of the hut, to longer focal lengths for recording detail of small objects such as the row of medicines on the shelf above Dr Wilson's bunk.

In wildlife photography in the Antarctic a high percentage of failures must be expected, but the really good results that one does get make it all so very worthwhile. Since for most people it is likely to be a once-in-a-lifetime trip, the cardinal rule is to take plenty of film and not to be afraid to use it; you can't nip back again to retake the bad shots.

General view of the interior of Scott's hut at Cape Evans, but the picture cannot capture the atmosphere that pervades this historic place.

Icicles on the underhang of a massive iceberg.

Overleaf *A low sun stains the sea red at Port Lockroy.*

Dorothy Hosking making friends with a group of King Penguins.

See p. 158 *Gentoo Penguin resting by whale vertebrae.*

Appendix I

Distribution of breeding birds of the Continental and Maritime Antarctic

	Continent	Antarctic Peninsula	Balleny Island	Peter I Øy	South Shetland	South Orkney	South Sandwich	Bouvet Øya
Emperor Penguin *Aptenodytes forsteri*	●	●						
Adelie Penguin *Pygoscelis adeliae*	●	●	●	●	●	●	●	●
Chinstrap Penguin *Pygoscelis antarctica*		●	●	●	●	●	●	●
Gentoo Penguin *Pygoscelis papua*		●			●	●	●	
Macaroni Penguin *Eudyptes chrysolophus*					●	●	●	●
Southern Giant Petrel *Macronectes giganteus*	●	●			●	●	●	●
Southern Fulmar *Fulmarus glacialoides*	●	●	?	●	●	●	●	●
Antarctic Petrel *Thalassoica antarctica*	●							
Cape Pigeon *Daption capense*	●	●	●		●	●	●	●
Snow Petrel *Pagodroma nivea*	●	●	●	?	●	●	●	●
Antarctic Prion *Pachyptila desolata*	?				●	●	●	?
Wilson's Storm Petrel *Oceanites oceanicus*	●	●	?	?	●	●	●	?
Black-bellied Storm Petrel *Fregetta tropica*					●	●		?
Antarctic Cormorant or Blue-eyed Shag *Phalacrocorax atriceps*		●			●	●	●	
Dominican Gull or Kelp Gull *Larus dominicanus*		●			●	●	●	?
Brown Skua *Catharacta lonnbergi*		●	?		●	●	●	●
South Polar or McCormick's Skua *Catharacta maccormicki*	●	●	●	●	● (few)	● (few)		
Antarctic Tern *Sterna vittata*	●	●		●	●	●	●	●
Wattled Sheathbill *Chionis alba*		●			●	●	?	
TOTALS								
Definite	10	15	5	5	17	17	14	9
Possible	1		3	2			1	3

Appendix 2

Distribution of breeding birds of the subantarctic islands

	South Georgia	Marion and Prince Edward	Îles Crozet	Îles Kerguelen	Heard and MacDonald	Macquarie Island
King Penguin *Aptenodytes patagonicus*	●	●	●	●	●	●
Chinstrap Penguin *Pygoscelis antarctica*	●				●	
Gentoo Penguin *Pygoscelis papua*	●	●	●	●	●	●
Rockhopper Penguin *Eudyptes crestatus*	● (few)	●	●	●	●	●
Macaroni Penguin *Eudyptes chrysolophus*	●	●	●	●	●	●
Wandering Albatross *Diomedea exulans*	●	●	●	●		●
Black-browed Albatross *Diomedea chlororhynchos*	●			●	●	●
Grey-headed Albatross *Diomedea chrysostoma*	●	●	●	?		●
Yellow-nosed Albatross *Diomedea chlororhynchos*		●		?		
Sooty Albatross *Phoebetria fusca*		●	●			
Light-mantled Sooty Albatross *Phoebetria palpebrata*	●	●	●	●	●	●
Northern Giant Petrel *Macronectes halli*	●	●	●	●	?	●
Southern Giant Petrel *Macronectes giganteus*	●	●	●	?	●	●
Cape Pigeon *Daption capense*	●		●	●	●	●
Snow Petrel *Pagodroma nivea*	●					
Narrow-billed Prion *Pachyptila belcheri*			●	●		
Antarctic or Dove Prion *Pachyptila desolata*	●		●	●	●	●
Salvin's Prion *Pachyptila salvini*		●	●			
Fulmar Prion *Pachyptila crassirostris*				?	●	
Fairy Prion *Pachyptila turtur*		●	●			
Blue Petrel *Halobaena caerulea*	●	●	●	●		●
Great-winged Petrel *Pterodroma macroptera*		●	?	●		
White-headed Petrel *Pterodroma lessoni*			●	●		●
Kerguelen Petrel *Pterodroma brevirostris*		●	●	●		

	South Georgia	Marion and Prince Edward	Îles Crozet	Îles Kerguelen	Heard and MacDonald	Macquarie Island
Soft-plumaged Petrel *Pterodroma mollis*		●	●	?		
White-chinned Petrel *Procellaria aequinoctialis*	●	●	●	●		
Grey Petrel *Procellaria cinerea*		●	●	●		●
Sooty Shearwater *Puffinus griseus*						●
Wilson's Storm Petrel *Oceanites oceanicus*	●		●	●	●	
Black-bellied Storm Petrel *Fregetta tropica*	●		●	●		
Grey-backed Storm Petrel *Garrodia nereis*	●		●	●		●
South Georgia Diving Petrel *Pelecanoides georgicus*	●	●	●	●	●	?
Kerguelen Diving Petrel *Pelecanoides exsul*	●	●	●	●	●	●
Antarctic or Blue-eyed Cormorant *Phalacrocorax atriceps*	●				●	
King Cormorant *Phalacrocorax a.albiventer*		●	●	?		●
Kerguelen Cormorant *Phalacrocorax albiventer verrucosus*				●		
Brown Skua *Catharacta lonnbergi*	●	●	●	●	●	●
Dominican Gull or Kelp Gull *Larus dominicanus*	●	●	●	●	●	●
Antarctic Tern *Sterna vittata*	●	●	●	●	●	●
Kerguelen Tern *Sterna virgata*		●	●	●		
Grey Duck *Anas superciliosa*						●
Yellow-billed Pintail *Anas georgica*	●					
Kerguelen Pintail *Anas acuta eatoni*			●	●		
Speckled Teal *Anas flavirostris*	● (few)					
Wattled Sheathbill *Chionis alba*	●					
Lesser Sheathbill *Chionis minor*		●	●	●	●	
South Georgia Pipit *Anthus antarcticus*	●					
Starling *Sturnus vulgaris*						●
Redpoll *Acanthis flammea*						●
TOTALS Definite	29	27	33	29	19	25
Possible			1	6	1	1

NOTE: Species deliberately introduced by man to the islands are excluded. Although the Starling and Redpoll were introduced to New Zealand, they have colonised Macquarie Island naturally.

Index

Acaena adscendens 48; *magellanica* 47, 48
Adelaide Island 37
Adelie coast 33
Land 120
Admiralty Range 22
Adventure 20
Africa 10, 16, 18
Agrostis magellanica 48
Aigle 18
Albatross: Black-browed – see *Diomedea melanophris*; Grey-headed – see *Diomedea chrysostoma*; Light-mantled Sooty – see *Phoebetria palpebrata*; Royal – see *Diomedea epomophora*; Shy or White-capped – see *Diomedea cauta*; Sooty – see *Phoebetria fusca*; Wandering – see *Diomedea exulans*; Yellow-nosed – see *Diomedea chlororhynchos*
Albatrosses 92–107
Alexander Island 23
algae, snow 40
Alopex lagopus 28
Amundsen, Roald 22, 25
Andes, South American 42
Annenkov Island 110
Antarctic 22, 23
Antarctic Circle 12, 18, 25, 30, 92; Continent 12, 14, 16, 18, 20, 23, 26, 28, 32–6, 38, 40, 42, 47, 59, 60, 64, 87, 91, 99, 110, 111, 115, 120, 125, 134; Convergence 12, 42, 44, 50, 52, 59, 115, 136, 140; Peninsula 22, 23, 26, 28, 32, 37, 38, 40, 52, 59, 60, 70, 120, 126
Antarctic Treaty 25
Antarctica 10, 16, 18, 22, 23, 25, 26, 91
Antarctica: Greater 33; Lesser 33
Anvers Island 120
Aptenodytes forsteri 33, 40, 59, 70, 76, 87, 89, 91; *patagonicus* 50, 57, 59, 60, 70, 75, 76, 80, 87
Arctic, The 26, 28, 38
Arctocephalus forsteri 136; *gazella* 41, 50, 116, 128, 131, 136, 140; *tropicalis* 50, 136, 140
Armstrong, Dr Terence 17
Australia 99
Azorella – see *Azorella selago*
Azorella selago 44, 48, 80

Balleny Islands 32, 33, 66
Bay of Whales 25
Bear, Polar – see *Ursus maritimus*
Belgica 23, 57
Bellingshausen, Thaddeus 20, 22
Bellingshausen Sea 22, 33
Beufort Island 66
Big Ben 44
BIOMASS 17
Bird Island 110, 115, 116, 120
Blechnum pennamarina 48
Bonner, Dr W.N. 17
Borchgrevink, Carsten E. 23
Bouvet Øya 18, 37, 41, 140
Bowers, Henry 25, 89
Bransfield, Edward 22
British Antarctic Survey 17, 60, 103, 131
Bullard Beach 70
Byrd Station 33

Cabbage, Kerguelen – see

Pringlea antiscorbutica; Macquarie Island – see *Stilbocarpa polaris*
Cape: Adare 22, 23, 52, 64, 70; Charles 23; Crozier 64, 70, 72, 87, 91, 131; Evans 25, 150; Hallett 64, 70; Royds 25, 64, 70
Catharacta lonnbergi 30, 116, 120; *maccormicki* 30, 33, 116, 120, 125; *skua* 30, 116
Cats 44, 47
Cecilia 22
Cherry-Garrard, Apsley 89
Chionis alba 40, 125, 126; *minor* 126
Colobanthus quitensis 38
Cook, Calotte Glaciaire 44
Cook, Captain James 18, 20, 94, 128
Cook Island 37
Coprosma pumila 47
Cormorant: Antarctic – see *Phalacrocorax atriceps*; Kerguelen – see *Phalacrocorax albiventer verrucosus*; King – see *Phalacrocorax a. albiventer*
Coronation Island 37
Coulman Island 70
Croxall, Dr J.P. 17

Daption capense 33, 109
Davis, Captain John 22
Deception Island 37, 59, 73, 146
Deschampsia antarctica 38, 47, 48
Diaz, Bartholomew 18
Diomedea: cauta 92; *chlororhynchos* 92, 101; *chrysostoma* 92, 101, 103, 107; *epomophora* 92; *exulans* 92, 97, 99, 101, 103, 107, 116; *melanophris* 92, 99, 101, 103
Dion Islands 40, 70
Discovery 22
Dufek Intrusion 16

Elephant Island 37, 66
Erebus 22
Eudyptes: chrysolophus 40, 44, 48, 50, 59, 70, 75, 76; *crestatus* 48, 59, 66, 75, 76; *schlegeli* 52, 57, 59, 70, 75, 76
Euphausia superba 73, 136
Evans, Edgar 25

Faeroe Islands 99
Falkland Islands 59
Festuca contracta 47
Forties, Roaring 44
Fox, Arctic – see *Alopex lagopus*
Fregetta tropica 40, 44
Fulmar, Southern – see *Fulmarus glacialoides*
Fulmarus glacialoides 33, 109, 115

Gallirallus australis 14, 47
Garrodia nereis 109
Gerlache Strait 22, 136
Gillham, Dr Mary 75
Gomeroy, Lt Adrien de Gerlache de 23
Gourlay Peninsula 66
Greenland 26
Gull, Dominican – see *Larus dominicanus*

Halobaena caerulea 109, 110
Hatch, Joseph 57
Heard Island 42, 44, 47, 48, 76, 80, 140
Hero 22
Hofmannfjella 110
Horlick Mountains 33
Hosking, Eric 16, 25, 142, 150
Hughes Bay 22
Humble Island 59
Hurd Point 70
Hut Point 25
Hydrobatidae 109
Hydrurga leptonyx 33, 41, 50, 126, 128, 131, 132

Île aux Cochons 44
Île de la Possession 44
Île de l'Est 44
Îles Crozet 42, 44, 47, 48, 50, 70, 75, 99, 115
Îles Kerguelen 42, 44, 47, 48, 50, 70, 99, 103, 136, 140
Institute, Scott Polar Research 17
International Geophysical Year (IGY) 25
International Whaling Commission (IWC) 25

Juncus scheuchzerioides 48

Kap Morris Jesup 26
Kildalkey Bay 70
King George Island 59, 64, 120, 125, 136
Knox Coast 33
Kristensen, Captain Leonard 22, 23

Larus dominicanus 40, 125, 126
Laurie Island 37, 64, 66
Laws, Dr R.M. 17, 131
Leptonochytes weddelli 33, 41, 50, 128, 131, 132, 134, 136
Lobodon carcinophaga 33, 13, 128, 131, 132, 136
Lozier, Jean Bouvet de 18

Macaroni Point 59
MacDonald Island 42, 44, 47, 48, 140
McMurdo Sound 33, 132
Macquarie Island 14, 42, 44, 47, 48, 50, 52, 57, 59, 70, 75, 76, 103, 109, 115, 128, 136, 140
Macronectes: giganteus 33, 44, 115, 116; *halli* 115, 116
Marguerite Bay 38
Marion Island 42, 44, 47, 50, 66, 70, 75, 76, 107, 115, 116, 126, 136
Maritime Antarctic 26, 28, 33–41, 42, 47, 59, 64, 126, 136
Mascarin, Pic du 44
Matthews, L. Harrison 92, 96
Mawson Peak 44
Mediterranean 30
Mice 44
Mirnyi 20
Mirounga leonina 41, 50, 120, 128, 131, 136
Montagu Island 37
Moseley, H.N. 107, 116, 126
Mount: Erebus 22; Faraway 110, 125; Hamilton 47; Jackson 37; Nivea 37; Ross 44; Terror 22
Murphy, Robert Cushman 72, 111, 115
Muskox – see *Ovibos moschatus*

New Zealand 14, 20, 57, 92
Nimrod 22
Nordenskjold, Otto 52
North Pole 26

Oates, Lawrence 25
Oceanites oceanicus 33, 44, 111
Omatophoca rossi 33, 128, 131, 134, 136
Orcadas 37
Orcinus orca 131, 132
Ovibos moschatus 28

Pachyptila: desolata 33, 40, 111, 115; *salvini* 115; *turtur* 115
Pagodroma nivea 33, 110, 111
Palmer: Archipelago 37; Land 22; Station 59, 131
Palmer, Nathaniel 22
Paul Beach 76
Pearson Point 110
Peary Land 26
Pelecanoides 44
Pelecanoides georgicus 44
Pelecanoididae 109
Penguin: Adelie – see *Pygoscelis adeliae*; Chinstrap – see *Pygoscelis antarctica*; Emperor – see *Aptenodytes forsteri*; Gentoo – see *Pygoscelis papua*; King – see *Aptenodytes patagonicus*; Macaroni – see *Eudyptes chrysolophus*; Rockhopper – see *Eudyptes crestatus*; Royal – see *Eudyptes schlegeli*
Penguins 52–91
Penguin-tussock – see *Poa hamiltoni*
Perseverance 128
Peter I Øy 32, 33
Petermann Island 59
Petrel: Antarctic – see *Thalassoica antarctica*; Black-bellied Storm – see *Fregetta tropica*; Blue – see *Halobaena caerulea*; Grey-backed Storm – see *Garrodia nereis*; Northern Giant – see *Macronectes halli*; Snow – see *Pagodroma nivea*; South Georgian Diving – see *Pelecanoides georgicus*; Southern Giant – see *Macronectes giganteus*; Wilson's Storm – see *Oceanites oceanicus*
Phalacrocorax: a.albiventer 126; *albiventer verrucosus* 126; *atriceps* 40, 126
Phoebetria: fusca 92, 101, 103, 107; *palpebrata* 99, 101, 103, 107
Pigeon, Cape – see *Daption capense*
Pleurophyllum hookeri 48
Poa: annua 44; *cookii* 44, 47, 48; *flabellata* 47; *foliosa* 48; *hamiltoni* 48
Pointe Geologie 111, 120
Possession Island 22, 52
Prasiola crispa 40
Pravda Coast 33
Prince, Dr P.A. 17
Prince Edward Island 42, 44, 47, 50, 66, 70, 75, 115
Princess Martha Land 22
Pringlea antiscorbutica 48
Prion: Antarctic – see *Pachyptila desolata*; Fairy – see *Pachyptila turtur*; Salvin's – see *Pachyptila salvini*
Procellariidae 109
Puffinus griseus 109
Pygoscelis: adeliae 33, 52, 59, 60, 64, 66, 70, 72, 73, 75, 91, 131; *antarctica* 33, 41, 59, 60, 66, 73, 75, 126; *papua* 40, 59, 60, 66, 73, 75, 107, 126

Queen Mary Coast 33
Queen Maud Coast 33
Queen Maud Land 22, 110
Queen Maud Mountains 33

Rabbit 47
Racovitza, M. 52
Rangifer tarandus 14, 42, 44
Rat: Black – see *Rattus rattus*; Brown – see *Rattus norvegicus*
Rattus: norvegicus 44; *rattus* 44
Reindeer – see *Rangifer tarandus*
Resolution 20
Rhacomitrum crispulum 48
Ronne Ice Shelf 22

Ross, Captain James Clark 10, 20, 52
Ross Island 22, 25, 33
Ross Sea 20, 22, 25, 64
Rostkovia magellanica 47

Saddle Island 66
Saxifraga oppositifolia 26
Saxifrage, Purple – see *Saxifraga oppositifolia*
Schmiracher Ponds 64
Scientific Committee on Antarctic Research (SCAR) 25
Scotia Arc 42, 50, 60
Scott, Robert Falcon 22, 25, 32
Scott Island 32
Seal: Antarctic Fur – see *Arctocephalus gazella*; Crabeater – see *Lobodon carcinophaga*; Kerguelen Fur – see *Artocephalus tropicalis*; Leopard – see *Hydrurga leptonyx*; New Zealand Fur – see *Arctocephalus forsteri*; Ross – see *Omatophoca rossi*; Southern Elephant – see *Mirounga leonina*; Weddell – see *Leptonychotes weddelli*
Seals 128–141
Sentinel Range 33
Shackleton, Sir Ernest Henry 22
Shag, Blue-eyed – see *Phalacrocorax atriceps*
Shearwater, Sooty – see *Puffinus griseus*
Sheathbill: Lesser – see *Chionis minor*; Wattled – see *Chionis alba*
Sheep 44
Shetland Isles 99
Shrub-burnet – see *Acaena magellanica*
Signy Island 37, 64, 66, 73, 136
Skua: Brown – see *Catharacta lonnbergi*; Great – see *Catharacta skua*; South Polar – see *Catharacta maccormicki*
Smith, Dr R.I. Lewis 17
Smith, William 128
Snow Hill Island 52
South Africa 99
South America 30, 99, 42, 136
South Georgia 14, 42, 47, 50, 57, 59, 60, 66, 70, 76, 99, 103, 109, 110, 115, 116, 120, 128, 131, 136, 140
South Orkney Islands 37, 40, 41, 42, 64, 66, 73, 136, 140
South Pole 16, 22, 25, 26, 32, 33, 91, 125
South Sandwich Islands 20, 37, 40, 41, 42, 140
South Shetland Islands 37, 38, 41, 42, 59, 64, 73, 120, 125, 128, 136, 140
Southern Cross 23
Southern Ocean 30, 109, 111, 128, 132
Soviet Antarctic Expedition 110
Spit Bay 80
Spitzbergen 99
Sterna vittata 33, 120
Stilbocarpa polaris 48
Stranger Point 64, 136
Subantarctic 26, 40, 41, 42–50, 92, 126, 136
Subtropical Convergence 115

Tern, Antarctic – see *Sterna vittata*
Terra australis incognita 10
Terra Nova expedition 22
Terror 22
Thalassoica antarctica 33, 110, 111
Theron Mountains 110, 125
Transantarctic Mountains 33
Tropic of Capricorn 92

Ui-te-Rangiora 18
Urusus maritimus 28

valleys, dry 33
Victoria Land 22, 33
Villers, Alan 20
Vindication Island 37
Vinson Massif 33
Vostock 20
Vostock Station 125

Weddell, James 128
Weddell Sea 16, 110
Weka – see *Gallirallus australis*
Whale, Killer – see *Orcinus orca*
Whalebirds – see *Pachyptila* species
Williams 22
Wilson, Edward 25, 52, 87
wind, katabatic 32

Zavodovski Island 66